Strategic Thinking in Action

A Practical Toolkit to Sharpen Your Strategic Mindset and Elevate Your Impact as a Leader

STEVEN HAINES

FOUNDER OF BUSINESS ACUMEN INSTITUTE

BΛI BUSINESS ACUMEN INSTITUTE

ISBN: 979-8-218-74951-4

CONTENTS

BUSINESS ACUMEN INSTITUTE

TRANSFORMING EMERGING LEADERS INTO STRATEGIC POWERHOUSES

In boardrooms across the globe, a critical gap is widening. Senior executives need emerging leaders who can navigate complexity, think strategically, and drive results. However, most professionals, across the spectrum of functions in many organizations, remain trapped in tactical thinking patterns that limit their impact and stunt their growth.

Business Acumen Institute stands at the forefront of solving this leadership development crisis. Through groundbreaking research and world-renowned training programs, we've cracked the code on what separates tactical executors from strategic leaders who shape organizational futures.

THE STRATEGIC THINKING IMPERATIVE

Today's business environment demands more than technical expertise and hard work. It requires leaders who can:

- **Think strategically** – See patterns others miss and connect decisions to long-term outcomes
- **Decide with data** – Transform information into strategic intelligence that drives competitive advantage
- **Act with agility** – Adapt quickly while maintaining strategic focus
- **Build believable business cases** – Translate insights into compelling arguments for resource allocation
- **Formulate actionable strategies** – Create plans that work in dynamic, uncertain environments
- **Fulfill the company's value proposition** – Align every decision with what creates lasting customer value

BEYOND BUSINESS ACUMEN: THE STRATEGIC THINKING BREAKTHROUGH

Our research reveals that while business acumen provides the foundation, **strategic thinking** has emerged as the critical differentiator. It's the cognitive glue that transforms capable managers into influential leaders who drive organizational success.

Business acumen is defined as *the portfolio of skills, behaviors, and capabilities needed to support an organization in achieving its financial and strategic goals.* **Strategic thinking** is what helps business acumen actionable under pressure, uncertainty, and complexity. Strategic thinking is the catalyst that transforms capable middle-level leaders into those who will influence the firm's future strategic and business success!

THE BUSINESS ACUMEN CANVAS: A COMPLETE LEADERSHIP DEVELOPMENT FRAMEWORK

Our multi-dimensional Business Acumen Canvas provides the comprehensive framework emerging leaders need to master every aspect of business impact. This model forms the foundation for our flagship, fully customizable workshop, Business Acumen Essentials.

Business Acumen Canvas

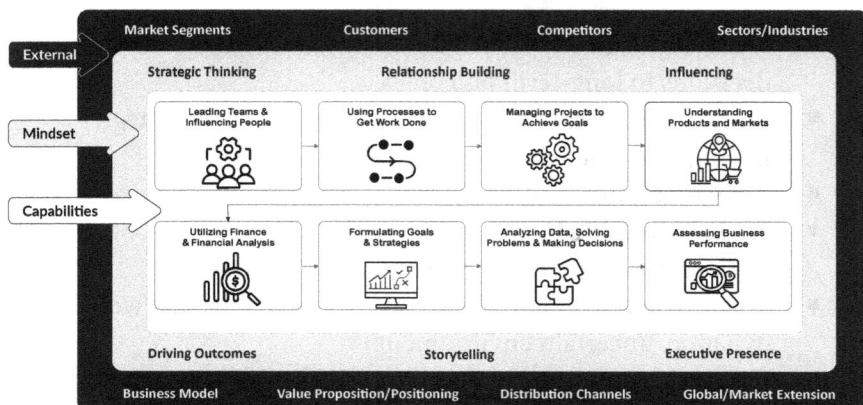

THE STRATEGIC THINKING MENTAL ARCHITECTURE: WHERE TRANSFORMATION HAPPENS

Through extensive research with thousands of professionals across industries, we've identified the precise cognitive architecture that separates strategic thinkers from everyone else. This breakthrough framework, the *Strategic Thinking Mental Architecture*, reveals the five interconnected capabilities that create compound leadership impact. It also serves as the backbone for the workshop entitled *Strategic Thinking in Action*.

Strategic Thinking Mental Architecture

5 Habits of Strategic Thinkers	Systems Thinking Lenses	Mindset Shifts	Problem Solving	Strategy Formulation
1. Ask better questions	1. Connection lens	1. I know to I'm learning	1. Step 0 – problem recognition	1. Beyond process to strategic intelligence
2. Recognize patterns	2. Feedback lens	2. Best practice to best for the situation	2. Cross-functional collaboration	2. Dynamic baseline
3. Zoom in – zoom out	3. Time lens	3. Control outcomes to influence conditions	3. Analyze root causes	3. State of the system
4. Learning mindset	4. Assumption lens		4. Make decisions	4. Work backwards
5. Time horizons	5. Leverage lens			5. AI pattern recognition

THE COMPETITIVE ADVANTAGE GAINED FROM STRATEGIC THINKING

Leaders who master these capabilities don't just perform with greater impact, they create compound advantages that accelerate over time. They spot opportunities others miss, solve problems others can't, and influence outcomes in ways that get them noticed for the right reasons.

Most importantly, they expand their field of view. Instead of reacting to events, they anticipate them. Instead of optimizing within constraints, they question the constraints themselves. Instead of competing in existing spaces, they recognize where new spaces are emerging.

FROM TACTICAL EXECUTION TO STRATEGIC IMPACT

Business Acumen Institute bridges the gap between where you are and where you need to be. We don't just teach skills, we transform how you think about complex challenges. After all, the leaders who shape the future don't just think differently, they think strategically.

To learn about our training and advisory services, write to me at sjhaines@business-acumen.com, or visit www.business-acumen.com

Steven Haines
Founder, Business Acumen Institute

WELCOME

"Think more strategically."

If you're reading this book, someone has probably suggested this to you. Maybe it was during a performance review, in feedback on a project, or in a conversation about your career development. If you're like most smart, capable professionals, your response was probably something like: "Okay, but what does that actually mean? And how do I do it?"

You're not alone. In my decades of working with managers and leaders across industries and around the world, I've heard this response from employees countless times. It's as if people know that strategic thinking is important, but they don't quite understand what it is, especially when they're running full speed on a treadmill that never seems to stop.

I get it. Early in my career, I too struggled with the concept of thinking strategically. I was lucky to have great mentors and coaches, though, who helped me figure it out through on-the-job experience. But many of you haven't been as fortunate, so I'd like to share what I've learned with you.

Here's what really gets me: I've watched brilliant leaders miss huge opportunities, solve the wrong problems, and make decisions that look great on paper but fail spectacularly in real business situations. Why? Because they were thinking tactically when they needed to think strategically.

The cost isn't just missed opportunities. It's teams that lose confidence when initiatives go nowhere; resources consumed on projects that don't create value for customers; careers that plateau because people can execute tasks brilliantly but can't see how their work connects to the bigger picture.

Here's what I've discovered: Strategic thinking isn't some mysterious talent that either you have or you don't. It's not about being the smartest person in the room or having access to better information. It's a learnable set of mental habits and perspectives that anyone can develop with the right guidance and practice.

Strategic Thinking in Action gives you that guidance in ways you may not have thought about . . . yet.

THE STRATEGIC THINKING GAP
(OR WHY NO ONE TAUGHT US HOW)

I continue to wrestle with the concern expressed to me by executives when I ask them about things that they feel may be missing in the development of emerging leaders and managers. They often say they regret that more people don't think strategically. Sometimes I suggest that this lack of strategic thinking is partially on them; that if they recognized it as a gap in training, then they could step in and coach those employees.

Most of us were never taught, as if it were a subject to teach, to think strategically, not in school, not in our companies. Instead, we learned about analytical frameworks and business processes. We became functional experts and got caught up in the tactical day-to-day business on our work treadmill. We needed help, but no one really knew how to find it.

What compounds this problem is that most resources (books, courses, etc.) on strategic thinking are either too abstract (theoretical frameworks you can't apply) or too specific (case studies from companies that are nothing like yours). You end up with concepts you can't use or examples you can't relate to.

That's the gap filled by this book.

THE ARCHITECTURE OF STRATEGIC THINKING

I'm going to be straight with you: this isn't another collection of strategy frameworks or abstract concepts. You won't find theoretical models that sound impressive but are impossible to apply in your daily work. Instead, I'll show you how to use a variety of thinking tools. The good news is that you don't need to do everything. You can try some and skip others. After this section, I'll provide you with a self-assessment that will offer you a perspective on where you are in your strategic thinking development. Once you know that, it's easier to identify where to go next on your journey as a strategic thinker.

One of the ways I'll offer you assistance is through the "Try This" exercises in most chapters. Think of these as practical experiments you can carry out to see what works for you. These are not hypothetical scenarios or artificial cases. They are realistic suggestions for you to use for situations you deem important.

Here are some areas that I cover in this book to help you be a more effective strategic thinker:

- See patterns others miss in the data and situations you encounter every day.
- Ask questions that reveal insights rather than just confirm what you already know.
- Recognize problems worth solving before they become crises.
- Design solutions that work in complex environments by training your brain to assess unexpected situations and spot opportunities others overlook.
- Navigate uncertainty with confidence by turning ambiguity from threat into strategic intelligence.

The difference is this: instead of teaching you what to think, this book teaches you how to think strategically about whatever you're facing.

STRATEGIC THINKING ISN'T STRATEGIC PLANNING

Here's something crucial to understand before you dive into the material: strategic thinking and strategic planning are not the same.

You can't formulate effective goals, strategies, and measurements if you don't think strategically first. You can't form insights from data about customers, competitors, finances, or operations without strategic thinking as your foundation.

The difference is crucial: Strategic planning creates documents and road maps. Strategic thinking promotes and builds capabilities. Strategic planning happens quarterly or annually. Strategic thinking happens every time you face a complex situation or must recalibrate your plans and realign your teams. Leaders who think strategically will likely have a greater impact on strategic planning and be more successful in the execution of those plans.

This book will help you master both the distinction and the connection.

HOW THIS BOOK WORKS

Strategic thinking isn't a single skill. Instead, it's a collection of thinking habits and mental models that work together. *Strategic Thinking in Action* builds your capability systematically. First, I'll group the chapters to shape the material. Later, I'll provide an abstract for each chapter.

Foundation (Chapters 1 and 2): Understanding the basics to thinking strategically and to developing the five core habits that separate strategic thinkers from everyone else. You'll start experimenting with these habits immediately.

Systems Awareness (Chapter 3): Learning to see the complex interactions and feedback loops that shape business outcomes. This is where you stop optimizing parts and start understanding things from a holistic standpoint.

Mindset Development (Chapter 4): Making three fundamental shifts: from knowing to learning, from best practices to contextual thinking, and from controlling outcomes to influencing conditions. These changes make everything else work better.

Real-World Application (Chapters 5 and 6): Transforming how you recognize and solve problems and how you develop strategies that create lasting competitive advantage.

Each chapter builds on the previous ones and includes practical experiments and suggestions. Think of them as workouts to fortify your strategic thinking muscle. Here are a few suggestions that might help you tackle this material:

1. Take the self-assessment so you have a baseline from which to formulate your learning goals.
2. Jot down some ideas about what you might want to have happen.
3. Use a notebook, paper or electronic, to take notes that are relevant to you. This approach may set off some ah-ha ideas for you to consider.

I don't want you to have to read the book and hope the light bulbs switch on. I'd like you to try to sensitize your mind so you're alert to situations that pop up and so you're better equipped to deal with a situation when it presents itself. To quote myself, you can't make an appointment with experience.

WHO THIS BOOK IS FOR

I'm a businessperson with a strong product management background. I apply customer segmentation techniques when I develop a product, and I've done so with this book. I did a lot of research and decided to write this book for people who are in middle to upper-middle management. You may fall into one of these segments:

- Emerging leaders who've been told to "think more strategically" but need practical guidance on what that means
- Experienced managers who want to elevate their impact beyond excellent execution
- Functional experts (marketing, sales, finance, operations, engineering, etc.) who need to see beyond their functional paradigms to understand how things connect
- Anyone who faces complex challenges without obvious solutions

If you've picked up a copy of this book, you've already demonstrated the intellectual curiosity and willingness to challenge your current thinking patterns. You've shown a willingness to try new ideas as you face unexpected challenges.

THE VALUE PROPOSITION FOR YOU

By the time you finish this book, you'll have developed strategic thinking capabilities that will serve you throughout your career. These include:

- *Pattern recognition skills* that help you see opportunities and threats before others do
- *Systems thinking ability* that reveals hidden dynamics shaping your business challenges
- *Mental flexibility* that allows you to adapt quickly as conditions change
- *Problem recognition capabilities* that help you spot the right problems early

- *Solution design approaches* that address root causes and create systemic improvements
- *Strategy development skills* that produce plans that actually work in dynamic environments

More importantly, you'll have the confidence that comes from knowing you can think your way through whatever challenges you encounter.

Here's how you'll know it's working: Within weeks, you'll catch yourself asking different questions in meetings. You'll spot patterns in data that seemed random before. You'll anticipate challenges your colleagues don't see coming. People will start asking how you saw something they missed.

This isn't theory. It's capability building that grows with practice.

YOUR STRATEGIC THINKING JOURNEY STARTS HERE

The cultivation, or improvement, of your strategic thinking requires that you think like a person who's training for a sporting event. You must practice purposefully. This means that you need a workout plan. As with any fitness plan, your commitment to the work will yield the best outcomes.

This is not a huge book. I've broken things down into small enough pieces that you can get your mind around things that resonate with you. As I suggested, start with the self-assessment and some attainable goals. Take small steps and then build on them. If you feel that one subject isn't relevant, skip over it. As my spin instructor says, you do you.

Here's the best part (at least, I think so): as you adapt and adopt, your behaviors will change. These behaviors are what's noticed by others. that are noticed by others. They will listen to the questions you ask, the problems you detect, the rationales you present. This will contribute greatly to how others perceive you as a leader.

The business world needs more strategic thinkers: leaders who can see beyond immediate crises to identify underlying patterns, who can connect the dots others miss, who can design solutions that strengthen over time.

You can be one of those people.

YOUR STRATEGIC-THINKING TRANSFORMATION IN SIX CHAPTERS

I've structured this book the way I'd teach it in person. Each chapter builds real capability you can use right away. You can read straight through or jump to whatever you need most. Either way, here's what awaits you.

CHAPTER 1: FUNDAMENTALS OF STRATEGIC THINKING

You've been told to "think more strategically," but nobody explained what that means or how to do it. Let's fix that. Strategic thinking isn't some mystical talent. It's pattern recognition combined with systems awareness, creative problem-solving, and the discipline to think across time horizons. You'll discover why strategic thinking feels so elusive (hint: most organizations accidentally discourage it) and learn the crucial difference between strategic thinking and strategic planning. In particular, you'll understand that strategic thinking is about continuous learning, not one-time planning exercises.

CHAPTER 2: FIVE HABITS TO BUILD YOUR STRATEGIC-THINKING MUSCLE

Now we get practical. Strategic thinking boils down to five learnable thinking habits that anyone can develop: asking better questions that reveal insights instead of just confirming what you already know, recognizing patterns and making connections across seemingly unrelated situations, fluidly zooming in on crucial details while maintaining a big-picture perspective, treating every situation as a learning opportunity, and thinking across time horizons to understand how today's decisions create tomorrow's reality. By the end of this chapter, you'll start thinking differently about problems you face every day.

CHAPTER 3: SYSTEMS THINKING: THE HIDDEN DISCIPLINE OF GREAT STRATEGISTS

Here's what nobody tells you: most problems that seem isolated are really systems problems in disguise. Fix one thing, break two others. Sound familiar? Optimize a process in your department and throw off a carefully crafted strategy. This chapter teaches you to see organizations as interconnected webs where every change creates ripple effects. You'll learn to peer through five systems-thinking lenses that reveal hidden dynamics: mapping connections, understanding feedback loops, recognizing time delays, surfacing assumptions that drive behavior, and finding leverage points where small changes create big impacts. Systems thinking prevents the expensive mistakes that come from optimizing parts while accidentally breaking the whole organization.

CHAPTER 4: MINDSET MATTERS: THE MENTAL MODELS THAT MAKE OR BREAK STRATEGIC THINKING

Your current mindset might be sabotaging your ability to think strategically without your realizing it. This chapter reveals three fundamental shifts that separate strategic thinkers from everyone else: moving from "I know" to "I'm learning," shifting from best practice to best for this situation, and evolving from controlling outcomes to influencing conditions. These aren't just philosophical concepts. They are practical approaches that make every strategic thinking tool more powerful. In this chapter I'll also offer you five specific techniques to rewire your thinking patterns.

CHAPTER 5: STRATEGIC PROBLEM-SOLVING: FROM RECOGNITION TO SMART DECISIONS

Most problem-solving approaches miss the most important step: recognizing problems worth solving before they become crises. I call this "step zero." It's a game-changing capability that lets you spot strategic issues while they're still manageable. You'll transform your approach from reactive "firefighting" to proactive solution design using a five-step process that works with complex, ambiguous challenges. Plus, you'll discover how modern tools like AI can enhance each step without replacing human judgment and learn to design solutions that address root causes while creating positive ripple effects.

CHAPTER 6: STRATEGIC THINKING MEETS STRATEGY FORMULATION: TRANSFORMING PROCESS INTO POWER

Strategy frameworks are everywhere, but most strategies still fail. Why? Because frameworks are only as powerful as the thinking that drives them. This final chapter shows how strategic thinking transforms proven strategy processes from template completion to strategic insight generation. Using five universal strategy formulation questions, you'll see how everything from previous chapters enhances every phase of strategy development. The result: strategies that work in dynamic reality, not just PowerPoint presentations. Not only that, you'll learn to build strategies through enhanced pattern recognition, systems analysis, and adaptive planning approaches that create sustainable competitive advantages even in uncertain environments.

ARE YOU READY TO BEGIN?

You don't have to master everything at once. Take your time. Like musicians who master a piece of music, you'll find that you'll be able to more easily finesse your own development as an evolving leader, which is why the next section involves your self-assessment. Think of it as the establishment of your strategic-thinking baseline so you can identify your most meaningful learning goals.

STRATEGIC THINKING SELF-ASSESSMENT

The self-assessment that follows is not a test. Consider it as your strategic thinking GPS. It will show you exactly where you are today and alert you to areas on which you might want to focus.

Whatever your starting point, you're about to begin a unique transformational journey that will change how you see challenges, make decisions, and create impact. The capabilities you'll develop don't just make you a better strategist; they make you a more effective leader, problem solver, and contributor to any organization.

Here's my advice to you: take the assessment honestly and really think about what you want to achieve. You don't need to do everything, but try to consider what you can focus on. Work with your manager to clarify your work plans so you can augment your annual goals. Keep good records of your accomplishments so you maintain the evidence of your work. After six to nine months, I recommend taking the assessment again to see how you've progressed. From this, you can recalibrate your goals. I believe that your future self as well as the organization in which you work will greatly benefit from the investment you're making in your evolution as a business strategist. Now, I'll describe the basics of the assessment.

There are six tables shown below; each is related to a chapter in the book. In each of those six tables, there are four activity statements. Place an X or similar mark in the cell that corresponds to the rating point you select. Then place the rating scale number in the rightmost column, add the scores, and place your total score for that table in the lower right cell. The rating scales are as follows:

1 = This rarely describes me.
2 = This sometimes describes me.
3 = This often describes me.
4 = This usually describes me.
5 = This consistently describes me.

Chapter 1: Fundamentals of Strategic Thinking

	Activity Statement	1	2	3	4	5	Your Score
1	I can explain the difference between strategic thinking and strategic planning.						
2	I regularly step back from immediate tasks to consider how they connect to broader business goals.						
3	I notice patterns or trends based on data or different situations that occur over time.						
4	I approach complex challenges by first understanding the whole system instead of individual parts.						
					Total Score		

Chapter 2: Five Habits to Build Your Strategic Thinking Muscle

	Activity Statement	1	2	3	4	5	Your Score
5	I ask questions that reveal new insights rather than just confirming what I already believe.						
6	I connect seemingly unrelated information from different areas of the business to gain new insights.						
7	I easily shift between detailed analysis and bigger-picture perspectives depending on the situation.						
8	I treat unexpected outcomes as learning opportunities rather than just problems to fix quickly.						
					Total Score		

Chapter 3: Systems Thinking: The Hidden Discipline of Great Strategists

	Activity Statement	1	2	3	4	5	Your Score
9	I map out how changes in one area might affect other areas before making decisions.						
10	I see how certain actions either reinforce problems or create positive momentum in my organization.						
11	I consider time delays between actions and results when evaluating the success of initiatives.						
12	I recognize when my own assumptions might be limiting my understanding of complex situations.						
					Total Score		

Chapter 4: Mindset Matters: The Mental Models That Make or Break Strategic Thinking

	Activity Statement	1	2	3	4	5	Your Score
13	I approach uncertain situations with curiosity because I view ambiguity as an opportunity to explore.						
14	I adapt solutions to fit specific contexts rather than applying standard practices universally.						
15	I focus my energy on influencing conditions I can shape rather than trying to control outcomes directly.						
16	I update my beliefs and approaches when new evidence contradicts my current thinking.						
					Total Score		

Chapter 5: Strategic Problem Solving: From Recognition to Smart Decisions

	Activity Statement	1	2	3	4	5	Your Score
17	I spot potential problems before they become urgent crises that demand immediate attention.						
18	I investigate underlying causes rather than just addressing symptoms when problems arise.						
19	I design solutions that address root issues while considering their broader impact on the organization.						
20	I build learning into my problem-solving process so that each challenge helps me anticipate future problems.						
					Total Score		

Chapter 6: Strategic Thinking Meets Strategy Formulation: Transforming Process into Power

	Activity Statement	1	2	3	4	5	Your Score
21	I analyze patterns in variances from plans instead of just examining single variances.						
22	I use a strategy formulation process, but I clarify desired outcomes before determining the best path.						
23	I synthesize insights from multiple data sources when formulating goals and strategies.						
24	I allow for flexibility in goal setting and strategy development to allow for conditions that might change.						
					Total Score		

Now that you've completed the assessment, it's time to understand what your responses reveal about your current strategic thinking capabilities. Your scores will help you identify both your strengths and your highest impact development opportunities. Remember, this isn't about judging where you are today; it's about creating clarity on where to focus your purposeful professional development efforts. Every strategic thinker started somewhere, and the most important step is knowing your baseline so you can chart your development path forward. First, use the table below to summarize your scores.

Summary Table

Chapter Reference	Chapter	Your Current Score	Your Desired Future Score
1	Fundamentals		
2	Five Habits		
3	Systems Thinking		
4	Mindset		
5	Problem-Solving		
6	Strategy Formulation		
Total Overall Score			

YOUR STRATEGIC THINKING LEVEL EXPLAINED

Tactical Expert (24–48 points)

You excel at execution and getting things done, but you're ready to expand your perspective beyond immediate tasks. You have strong functional expertise and probably want to connect your work to provide a more significant impact to the business. You're skilled at solving problems as they arise but would benefit from adjusting your mindset to anticipating problems before they become urgent priorities.

Your possible development focus: Building strategic awareness and pattern recognition

Emerging Strategist (49–72 points)

You're developing strategic awareness and starting to see patterns beyond your immediate responsibilities. You ask good questions and understand that context matters, but you want to strengthen your ability to think systemically. You're beginning to balance short-term execution with longer-term thinking and are ready to take your strategic capabilities to the next level.

Your possible development focus: Strengthening systems thinking and strategic problem-solving

Systems Thinker (73–96 points)

You naturally see connections and understand how different parts of the business affect each other. You're comfortable with complexity and uncertainty, and you regularly contribute strategic insights to your team. You want to refine your strategic thinking tools and apply them more systematically to create even greater impact.

Your possible development focus: Mastering strategic mindset and advanced problem-solving

Strategic Leader (97–120 points)

You consistently demonstrate advanced strategic thinking across multiple situations and time horizons. Others seek your perspective on complex challenges, and you regularly influence strategic direction in your organization. You're looking to deepen your mastery and help others develop their strategic-thinking capabilities.

Your development focus: Mastery refinement and organizational capability building. Use your skills to coach and develop others.

1

FUNDAMENTALS OF STRATEGIC THINKING

Key Points
- The biggest barrier to strategic thinking is organizational culture, not individual capability.
- Strategic thinking happens in real time, not during annual planning sessions.
- You'll know you're thinking strategically when you start seeing connections that others miss.

Knowing how to think empowers you far beyond those who know only what to think.

—NEIL DEGRASSE TYSON

Two smart, capable professionals sat across from me, shuffling papers and avoiding eye contact. We'd just finished a strategic planning workshop their boss called "invaluable," but now they had a problem.

"She wants us to think more strategically," one finally said.

I paused. "What does strategic thinking mean to you?"

The ceiling suddenly became very interesting to both of them.

WHY WRITE ABOUT STRATEGIC THINKING?

That moment with those two frustrated professionals stayed with me for weeks. Here were talented, competent people who had just completed a workshop on strategic planning, yet they couldn't define the topic even when their boss demanded it of them.

So I did what any curious businessperson would do. I researched. I downloaded books, studied academic papers, talked to executives. What I found was disappointing: most resources were filled with either abstract theories you couldn't apply or case studies from companies nothing like yours.

But the real eye-opener came from my conversations with leaders in my client companies. When I asked them what "thinking more strategically" meant, their responses varied dramatically. One CEO emphasized long-term planning. Another focused on competitive analysis. A third insisted it was about innovation and creative problem-solving. One even said something like "I don't see it, but I know it when I see it."

This inconsistency is more than frustrating. To me, I think it's dangerous. In organizations where efficiency and speed determine competitive advantage, unclear definitions kill collaboration. When teams don't share a common understanding of strategic thinking, their efforts become sub-optimal, even counterproductive.

Whether we're talking about business acumen, leadership, strategy, or strategic thinking, each of these terms means different things to different people. Without clear, consistent definitions, you can't develop these capabilities systematically, and you certainly can't apply them effectively when it matters most.

That's why I wrote this book.

A PRACTICAL DEFINITION FOR STRATEGIC THINKING

If your goal is to become a more proficient strategic thinker, you've found the right source for ideas and inspiration. It's my belief that I can help you become more sensitive to the opportunities that can emerge when you think more strategically. As with my other books, I intend to use this as a platform from which you can build this core capability, using practical tools and applicable guidance.

Let's dive into what makes strategic thinking so important and how you can improve this multifaceted skill set. Basically, strategic thinking is the ability to see patterns, understand connections, and make decisions that create lasting value in complex, changing environments. More specifically,

Strategic thinking is a mental process used to make sense of dynamic data, disparate observations, and other information to expose linkages or patterns, establish meaningful goals, determine appropriate courses of action, and continually refine implementation through metric-driven feedback, all operating across different time horizons.

I realize that this is a complex definition and goes beyond what others might use to define strategic thinking. However, my goal is to foster unique perspectives that may challenge your current thinking. Consider reflecting on a time when you encountered a complex situation and made a suboptimal decision. Sometimes you don't have to decide on an action right away. You may need to step away from your day-to-day blocking and tackling and other tactical activities to view things from different vantage points. Like an army general, you need to assess the battlefield and a host of different variables before sending your troops into battle. Military commanders are usually effective strategic thinkers.

STRATEGIC THINKING AS A DYNAMIC PROCESS

One of the biggest misconceptions about strategic thinking is that it's a fixed characteristic, something you either have or don't have, or something you're supposed to do when you're told to do so. This is a flawed perspective. Strategic thinking isn't a single moment of inspiration, nor is it a one-time event, such as a brainstorming session or a planning workshop. Rather, it's an ongoing process underscored by purposeful practice.

It's something you develop, nurture, and refine over time. Figure 1.1 shows strategic thinking as a process.

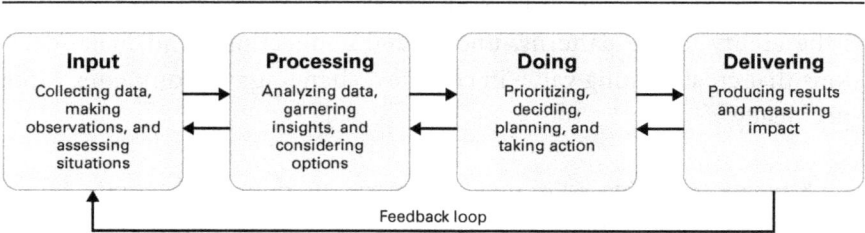

Figure 1.1 Strategic Thinking Process

Input	Processing	Doing	Delivering
Collecting data, making observations, and assessing situations	Analyzing data, garnering insights, and considering options	Prioritizing, deciding, planning, and taking action	Producing results and measuring impact

Feedback loop

Strategic thinking has a lot of mental back-and-forth processing, and it's far from linear, even though this is how it appears in the diagram. You know the complexities of the world and the dynamism of business, so it stands to reason that strategic thinking isn't static or episodic. If people tell you that you must be a strategic thinker and can just turn it on at will, they're missing the point, and you'll be frustrated. The world doesn't sit still, and neither should your approach to strategic thinking.

Like the dynamism of the strategy formulation process, strategic thinking is iterative. In other words, you're constantly cycling through analysis and refinement. You may view your company's strategic-planning process through the lens of timelines, activities, deliverables, and results. However, in reality, strategic thinking is far from linear. The best strategic planners are, by association, strategic thinkers. They constantly examine data, ask questions, probe for meaning, and consider the next steps. Like any executive, your challenge is to tune into market trends, operational activities, and financial results to recognize business patterns that help you draw conclusions that may not be obvious.

THE POWER OF PATTERN RECOGNITION

One of the key aspects of the strategic thinking process is pattern recognition. You must be able to make sensible, simple abstractions from a complex web of hard data and observations and connect the dots between seemingly unrelated data points. Like the work of any practitioner (e.g., surgeons, lawyers, and musicians), you get better with purposeful practice.

Improvement of this process is directly related to your evolving proficiency. The process seems linear, but it is truly iterative. In business your company serves customers. Over time their needs and preferences change. Competitors do unexpected things. The economy may fluctuate. Technologies will evolve. Therefore, strategic thinkers know that data are dynamic and market conditions fluid. As you evolve in your career, you'll realize that you will need to adjust your perspectives and refine your thinking. This is why the most effective leaders are not just strategic thinkers; they question and they learn.

STRATEGIC THINKING IN ACTION: A REAL-WORLD EXAMPLE

Let me show you what this looks like in practice. I once worked with the leadership team of a complex, multiline software company. They were concerned that the product teams were not delivering the promised value proposition to their customer base. We asked them to share their product strategies and road maps with our team. We also interviewed many who worked in product management, product development, and marketing. The information they shared was helpful, and they were all passionate about the features they were delivering.

However, what we ultimately learned was that their strategies didn't consider the vast market changes that we observed, including an impending economic downturn. Their focus was almost entirely internal. They were more interested in getting features incorporated into their products to make specific release dates and meeting targets on a tactical roadmap. But in doing so, they had taken their eye off the customer. Their internal focus took their attention away from core customer assumptions that had been made many months before.

When we asked them to begin with an analysis of customer sentiment data, they realized they had to conduct fresh research. When they did, their insights revealed a clear pattern: customer needs had shifted. Budgets were tightening, and leaders in customer companies were prioritizing efficiency and cost savings over feature expansion. I spoke to a few of their customers, who, and I'll paraphrase here, said they were continually bombarded with new features and user experience (UX) design changes that they found disruptive and couldn't possibly absorb. The software company's designers and developers were locked into a product roadmap that had been done months before, and they were just working off the backlog (list of features to develop). They had no clue that their assumptions were out of date and that their value proposition was no longer accurate. Other departments just played along. Salespeople tried to close deals but were failing more often. Marketing people were launching campaigns that had little response. In essence, it was the beginning of a downward spiral. In the end, the company's strategic plans, while well-intentioned, were built on yesterday's reality, not today's. Not only that, but its leaders had failed to anticipate what might happen next.

This is where the iterative nature of strategic thinking becomes essential. *Market conditions don't wait for your next planning cycle.* Competitive landscapes shift unexpectedly. The economic environment may change in a heartbeat. The most effective leaders recognize that strategy is not a set-it-and-forget-it exercise. Strategic thinkers learn, adapt, and refine their perspectives continuously, rather than blindly executing a plan that may no longer be relevant.

LEARNING ABOUT SYSTEMS TO UNDERSTAND STRATEGIC THINKING

Have you ever heard the term *ecosystem*? It's used to describe any group of organisms coexisting within the natural environment. In our world, social and business ecosystems allow us to visualize how things are connected or coexist. This offers us options for how to deal with complex situations. By association, a system is a collection of interconnected parts that work together to achieve an outcome or purpose.

Here's how the ecosystem of business begins to take shape. In Figure 1.2, note the four basic functions. Consider these as subsystems of a larger system.

Figure 1.2 Business Functions as Subsystems

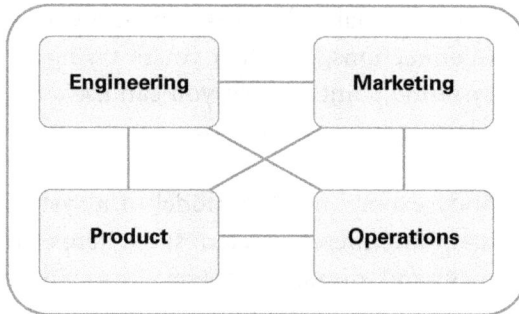

You'll see that there are six connection points between the four departments. If there were six departments, there would be fifteen connection points. This simple visual is relatable because it's what most of us are familiar with. However, with six connection points, if the work done by one department is unexpected or not synchronized with the company's strategy, there are reverberations felt across the entire enterprise. These "ripple effects" can affect not only internal efficiency but also customers, suppliers, and others in the business ecosystem. Figure 1.3 shows the relationships among the different entities of a business ecosystem.

Figure 1.3 Business Ecosystem

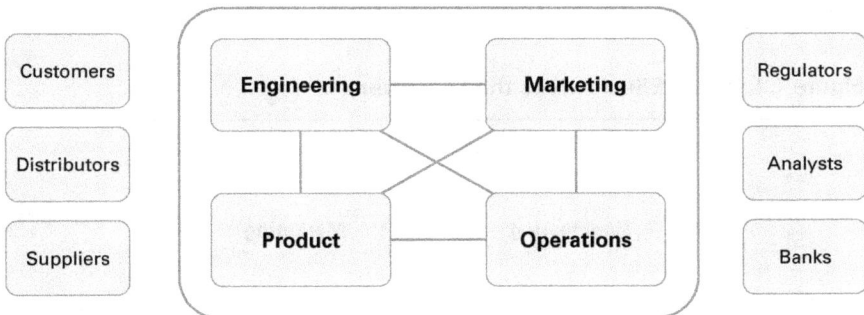

The business ecosystem not only connects functional departments to one another; it connects the organization to the outside world. While there are many more external influences, I'm sure you get the idea that there are a lot of connections, so when you're trying to assess a given situation from a systemic point of view, you can use a diagram or visual tool to map how data or information flows (or doesn't) to assess what's going on.

The human body exemplifies the model of a system because it's a dynamic metasystem of interconnected subsystems. The circulatory, nervous, respiratory, and digestive systems may appear to function independently, but they are profoundly interdependent. When infection threatens the body, for example, a fever that develops as a result isn't merely an isolated event but a coordinated systemic response, demonstrating how components work collaboratively to protect the whole.

Similarly, strategic thinking demands we look beyond individual elements to understand their interactions. For example, let's say the marketing department made erroneous assumptions about emergent customer needs. The ripple effects would be felt by the product group, who relied on the information for their strategy and road map, and the engineering group, who relied on the associated use cases for their designs and specifications. Ultimately, this would impact the customers, who would not get what they needed, and the company's competitive position would be imperiled. This ripple effect is shown in Figure 1.4, which shows how one error can reverberate across the entire ecosystem. Here's why this matters so much.

Figure 1.4 Ripple Effect Across the Ecosystem

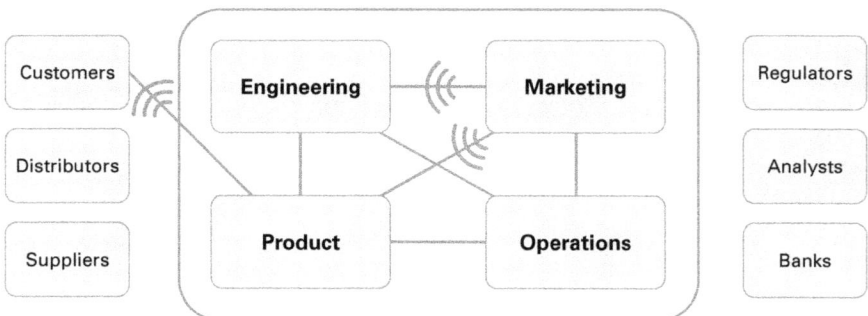

By embracing a systems perspective, you can anticipate ripple effects, identify powerful leverage points, and make decisions that optimize the entire organization rather than just isolated departments. This ultimately creates competitive advantages that wouldn't be possible through fragmented thinking. Chapter 3 is devoted to this important topic.

One of the most important enablers of systems thinking is the use of a feedback loop. A feedback loop simply refers to a cycle where information circles back to influence future actions. Why? Because systems are not static; they evolve based on inputs, activities, decisions, and outputs. A feedback loop helps strategic thinkers anticipate how a change in one area will impact the broader system over time. You'll note in Figure 1.1 that there's an arrow that connects the rightmost box (Delivering) to the leftmost box (Input). That's a feedback loop and is vital because it enables you to steer the business (or your part of the business) just as you'd drive a car and make corrections as conditions arise.

Consider a company that aggressively discounts its products to drive short-term sales. This sometimes happens in the automobile industry. When inventories build, dealers roll out the incentives. Initially, revenue spikes, but over time, customers begin to expect lower prices, eroding brand value and pricing power. Worse, competitors may respond by slashing their prices, triggering a race to the bottom. This is an example of a reinforcing feedback loop, in which an action creates conditions that intensify its own effects, often in an unintended or negative way. Strategic thinkers with a systems viewpoint look beyond the immediate impact of a decision and evaluate how it might amplify or disrupt other parts of the system.

Here's another benefit of systems thinking. When people take a linear approach to strategy, they tend to believe that cause and effect are simple and predictable. But real-world business challenges are not that straightforward. When people don't assume a systems-thinking approach, they often find themselves in a reactive mode.

A classic example of this is corporate "reorg," or reorganization. Leadership may decide to reorganize departments to streamline efficiency, but if they fail to consider the cultural and operational ramifications, they may inadvertently create new inefficiencies, lower morale, or disrupt customer relationships. Strategic thinkers anticipate second- and third-order effects before making decisions, ensuring that adjustments lead to long-term gains rather than short-term fixes.

THE ART OF ZOOMING IN AND ZOOMING OUT

Great strategic thinkers have the ability to zoom in and zoom out where they can shift focus between specific details and the bigger picture. This allows them to see the micro-level details while maintaining awareness of the macro-level system. They recognize that a single data point, a departmental goal, or a market trend is just one part of a larger, interconnected network.

Take the example of supply chain management. Strategic leaders don't just focus on optimizing internal logistics; they also zoom out to assess geopolitical risks, trade policies, and emerging technologies that might reshape the industry. Then they zoom back in to determine how their organization should adapt, whether through supplier diversification, process automation, or strategic partnerships. I'll provide more information on this topic in Chapter 3.

TRY THIS:
Pick a recent decision you made. Walk through the five elements of the mental architecture. Which elements did you use well, and which did you skip?"

BASIC ELEMENTS THAT INFLUENCE STRATEGIC THINKING

Strategic thinking usually involves perspectives gathered across multiple time horizons. A way to view this is through four lenses: past, present, potential, and possible path. As I've mentioned, business is dynamic, and unexpected things happen. That's why strategic thinkers are constantly reviewing data, including

- Financial data (revenue, expenses, and profit)
- Customer data (segments, purchases, and preferences)
- Competitor data (companies, products, and market share)
- Industry or market data (economics, regulations, technology, etc.)
- Internal operating data (service, support, development, etc.)

Think of it this way: multiple time horizons allow you to simultaneously consider immediate, near-term, and longer-term consequences of possible goals, options, and outcomes. This temporal flexibility is augmented by an inherent adaptability, because strategic thinkers don't operate within rigid boundaries; they have flexible perspectives that evolve as circumstances change and new data come to light.

THE INNOVATION CONNECTION

The strategic thinking capabilities I've described: pattern recognition, systems awareness, and creative problem-solving, naturally lead to innovation. When you can see patterns that others miss and understand how systems connect, you're positioned to envision possibilities that aren't immediately obvious to others.

Consider how this works in practice. When you apply pattern recognition across different industries or domains, you might spot a solution from healthcare that could transform manufacturing, or a retail approach that could revolutionize financial services. When you use systems thinking to understand how different parts of a business connect, you can identify intervention points where small changes create breakthrough innovations.

This is exactly what made leaders like Steve Jobs and Bill Gates so effective. Jobs didn't just create products; he recognized patterns in human behavior that others missed and designed systems that elegantly solved unspoken needs. Gates saw patterns in computing power growth and connected them to his vision of personal computers in every home, a systems-level insight that seemed impossible to most people at the time.

Here's what I want you to understand: innovation isn't just creativity; it's strategic thinking applied to opportunity recognition. When you train yourself to look at business or financial data differently, spend more time understanding your customers' evolving needs, or work more closely with people in different departments, you're building the foundation for innovative thinking.

This systematic approach to innovation builds on everything I've shared with you. The strategic thinking process from Figure 1.1 becomes your innovation engine: gathering diverse inputs, processing patterns across domains, synthesizing insights, and delivering novel solutions that create lasting value.

DEVELOPING YOUR STRATEGIC THINKING CAPACITY

In my own career, I was always curious about how things worked and how different people did their jobs. I liked visiting customers and learning their businesses. I've always been an enthusiastic consumer of business news. I have a passionate hunger for information to feed my brain.

There's another thing that I learned about my own ability as an evolving strategic thinker: I found that when I wanted to cut out the noise of life and let my thoughts run free, I would meditate or go to the gym. I'm an avid indoor cyclist (yes, I love spin classes), and I find that when my mind goes adrift with the music and the exercise, I get really creative. (The idea for this book came to me during a SoulCycle class.)

Consider this approach to nurturing creativity: choose your own ways to clear away tension, take a break from the action of the day, and let your mind wander to explore new perspectives and ideas. This mental space allows you to look beyond incremental improvements in your own organization to actions that may be more transformative. While there's no one-size-fits-all formula for developing strategic thinking and creativity, the structured practices in this book combined with your deliberate application will help your brain start making new connections.

WHERE DOES CRITICAL THINKING FIT?

When I took AP English in high school, I tackled James Joyce's *Portrait of the Artist as a Young Man*. It awakened my critical thinking abilities before I even had the vocabulary to name what I was learning. For example, throughout the book, Joyce uses red and green imagery in key moments. In a paper I wrote on my reactions to the book, these references seemed purely descriptive to me. My mom, an English major, took an interest in my paper. After reading my first draft, she asked me, "Why red and green?" I gave a shallow answer about Ireland's green landscapes and didn't think much more of it. She challenged me to dig deeper, so I headed to the library for some research.

I discovered that Joyce often used the red-green color contrast deliberately. Green evoked Ireland's lush natural beauty, but also renewal and the hope of a different life. Red was often tied to sin, temptation, or the constraints of religion and social order. This juxtaposition revealed the protagonist Stephen Dedalus's internal conflict and his gradual rejection of religious dogma in favor of artistic freedom. Through this lens, Joyce's color palette became a powerful way to explore the tension between tradition and independence.

That experience taught me to dissect language and imagery to uncover underlying tensions and contradictions. It fundamentally changed how I approached information and, ultimately, data analysis in business. I learned to examine not just what was presented but also the complex, often conflicting values and systems beneath the surface.

When it comes to critical thinking, it's about being aware of how you're thinking while you're in the middle of a specific situation. Richard W. Paul, who started the Foundation for Critical Thinking, puts it nicely: "Critical thinking is thinking about your thinking while you're thinking in order to make your thinking better." In plain terms, it's an ongoing process in which you gather information, put it together, and analyze it to make better decisions. That part about "thinking about your thinking" is just recognizing when your own biases or assumptions may be getting in the way. When you think critically, you're trying to get to the root of problems, consider different viewpoints, weigh your options, then make a choice. Leaders use this skill every day for routine challenges and major opportunities alike. In sum, critical thinking is an important aspect of strategic thinking, just as creative thinking is.

THE ROLE OF CREATIVE THINKING

I've explained what I believe is important about critical thinking and its role in strategic thinking. From another vantage point, how do we generate unique ideas or come up with innovative ways to solve problems in business? This is where creative thinking enters the strategic thinking equation. While critical thinking helps us analyze what exists, creative thinking enables us to envision what could be. This complementary cognitive skill works alongside the systems thinking I discussed earlier. It allows us to not only understand complex interconnections but also reimagine them entirely. Together it provides us with a unique ability to generate novel ideas and envision innovative solutions.

While I'm on the topic of innovation, I don't want you to be confused by a frequently used term in business: the *innovation process* (the structured stages companies use to move from idea to market). Many executives, especially those who are more technically inclined, tend to lean on their innovation process to come up with creative product ideas. The problem as I see it is that their innovation process is structured, linear, and designed to transform creative ideas into new or enhanced products. To me, this is confusing because you can't "engineer" creativity. When I examine product failures in working with my clients, I often learn that process centricity (rigid adherence to steps over adaptive thinking) without data and creative thinking was the cause.

As I see it, creative thinking is about seeing possibilities where others don't. Unlike the structured innovation processes many companies rely on, real creative thinking can't be forced through a predefined set of steps and process diagrams. This is exactly why industrial design firms such as IDEO and Frog Design have been so successful. To my understanding, they've built their entire approach around human-centered creative problem-solving rather than rigid processes. They understand that creativity requires space for imagination, unexpected connections, and, sometimes, productive detours.

I've had the privilege to visit with people who work in these types of industrial design firms, and I noticed they are not constrained by rigid linear processes. They are like kids I've known who colored outside the lines in coloring books; they weren't constrained by the heavy lines of the picture but instead saw things differently.

If I could summarize what I'd like you to take away from this, I'd offer this suggestion: if you can pull back from your day-to-day life on the business treadmill and get out of the office to spend more time with customers, or study new technologies, or learn things outside of your current field of interest, you might find flashes of inspiration that could help you think more creatively and improve your overall strategic thinking acumen.

COMMON BARRIERS TO STRATEGIC THINKING

Organizations inadvertently create obstacles to strategic thinking through their structures and processes, but you can learn to navigate and overcome these barriers systematically.

Organizational Barriers and How to Navigate Them

Quarterly reporting cycles often drive short-term activities at the expense of long-term vision. To counter this, protect time for strategic reflection by scheduling it like any other important meeting. Block 30 minutes weekly for "strategic thinking time" and use it to step back from immediate pressures to consider longer-term patterns and implications.

Departmental silos prevent the cross-pollination of ideas necessary for strategic perspectives. Build relationships across functions by scheduling monthly conversations with colleagues in other departments. Ask them about their biggest challenges and look for connections to your own work. This cross-functional perspective becomes invaluable for systems thinking.

Performance metrics that focus exclusively on immediate results can discourage exploration of uncertain but potentially valuable opportunities. When this happens, start small. Propose pilot projects or limited experiments that let you test strategic ideas without risking major resources or commitments.

Personal Barriers and Mental Traps

Your own mind can harbor cognitive biases that impair strategic thinking. *Confirmation bias* leads you to seek information that supports existing beliefs while filtering out contradictory evidence. Combat this by actively seeking out perspectives that challenge your assumptions. Ask yourself: "What would someone who disagrees with me say about this?" *Recency bias* gives disproportionate weight to recent events over historical patterns. Before making decisions based on recent data, ask: "Is this part of a longer-term pattern, or is it an anomaly that's getting too much attention because it just happened?" *Status quo bias* creates resistance to change even when current approaches are demonstrably failing. When you catch yourself defending "the way things are," pause and ask: "What evidence would convince me that a different approach might work better?"

Your Strategic Thinking Defense System

Overcoming these barriers requires building what I call your strategic thinking defense system. Here's how:

1. Develop awareness by recognizing when you're falling into these traps.
2. Create environmental supports like protected thinking time and diverse perspectives
3. Build the habit of asking "why" questions multiple times to uncover deeper patterns beneath surface-level observations.

Remember: awareness of these biases is the first step toward mitigating their impact. The goal isn't to eliminate them completely. That's impossible. The goal is to recognize when they might be influencing your thinking and have tools ready to counter their effects.

SUMMARY

Strategic thinking is not an innate talent that some people have and others don't. Instead, it is a learnable, multifaceted process that combines several approaches to make sense of business complexity. It can be enhanced through purposeful learning. Here are some points to remember:

1. *Process orientation.* Strategic thinking is ongoing, dynamic, and iterative. It is not confined to periodic planning (e.g., strategic planning calendars) that tend to be the norm in most organizations.
2. *Pattern recognition.* The ability to connect seemingly unrelated data points across the company, including data about customers, financials, operations, and market trends, can help you form meaningful insights.
3. *Systems understanding.* You can adjust your perspectives when you recognize that a business is made up of dynamic, interconnected systems and subsystems where changes in one area create ripple effects across the organization.

4. *Critical analysis.* When you can examine a situation and question assumptions or challenge conventional methods, you'll be able to uncover deeper meaning or contradictions. Alternatively, you may find areas that can be favorably reinforced.

5. *Creativity counts.* In strategy formulation, there's an element that includes a vision for the future. Vision cannot be dictated, but if you can consider possibilities or future solutions that are outside the field of view of others, you may have a greater impact on the organization's success.

6. *Balance matters.* When you can zoom in to understand important details, then zoom out to see the bigger picture, your perspectives will be enhanced. When you're able to demonstrate this capability to others or if you can teach this to others, your value as a strategic leader will be enhanced.

As you continue reading, try to reflect on where you currently stand along the strategic thinking continuum and how you might apply these principles to view familiar situations through a new lens. Understanding these fundamentals provides the foundation, but transformation happens through consistent practice.

In the chapters that follow, I'll move from theory to action. In the next chapter, I'll explore five specfic habits that separate strategic thinkers from everyone else. In the chapters that follow, I'll show you practices you can begin implementing immediately with whatever challenges you're currently facing. Remember that becoming a strategic thinker is a journey of improvement. It requires both disciplined practice and the courage to challenge your own thinking.

2

FIVE HABITS TO BUILD YOUR STRATEGIC THINKING MUSCLE

Key Points
- Better questions reveal insights instead of confirming assumptions.
- Pattern recognition across domains helps you spot opportunities that others miss.
- Thinking across time horizons prevents short-term decisions from creating long-term problems.

Watch your thoughts, they become actions. Watch your actions, they become habits. Watch your habits, they become character."

—LAO TZU

The leaders who seem most "naturally strategic" aren't necessarily smarter or more experienced than you. They've just trained themselves to think differently in five specific ways.

Most people assume that strategic thinking is a mysterious talent you either have or don't. But I've watched mid-level managers transform into strategic leaders by mastering five simple habits, habits so practical you can start using them today.

Let me be clear about what this chapter will and won't give you. This isn't about learning more theories or abstract concepts. The world doesn't run on theories. It works on the real, sometimes messy discipline of developing habits and mindsets that help you see further, connect more dots, and make better decisions. My aim is to give you concrete tools and advice you can use right away, regardless of your role or level of experience.

WHY STRATEGIC THINKING FEELS SO ELUSIVE

Right now, I want to speak directly to you, not as an author and teacher, but as someone who's been in the corporate trenches, faced uncertainty, and made tough decisions. Perhaps you're a manager who feels overwhelming pressure to deliver results when market conditions are changing. Maybe you're leading a team, experiencing shifting priorities and increasing demands. Alternatively, you could be leading a cross-functional team and attempting to influence outcomes beyond your immediate control.

No matter what situation you find yourself in, it's my belief that your ability to think strategically is what can set you apart from others who merely react.

I see this constantly in my training and longer-term applied learning programs with smart, capable professionals who get trapped in cycles of reactivity. They go from meeting to meeting and crisis to crisis. They don't have the mental space needed to pull back and consider strategic issues. Crises, however, often can be averted if the time is taken to view situations from a strategic perspective.

In my conversations with managers, aspiring leaders, and executives from a wide range of industries, I've heard a common refrain: "I know strategic thinking is important, but I'm not sure how to actually do it."

It's for this reason that I developed a framework of five practical habits that separate strategic thinkers from everyone else. These habits aren't tied to a job title or business function. They're mental muscles that anyone can build. Like physical fitness, strategic thinking requires intention, repetition, and reflection. When you master these five habits, I know that you will be able to build your strategic thinking muscle that will become a vital part of how you lead, decide, and act every day. To start, consider Figure 2.1, which depicts the five habits of strategic thinkers.

Figure 2.1 The Five Habits of Strategic Thinkers

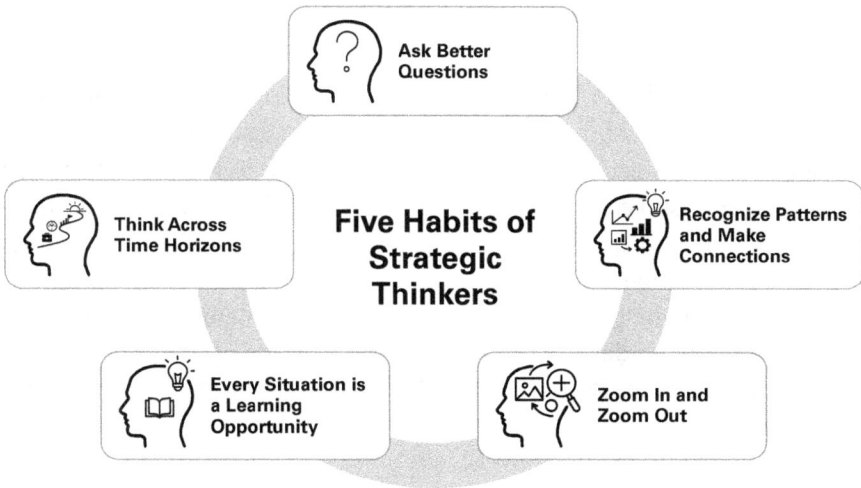

THE FIVE LEARNABLE STRATEGIC THINKING HABITS

Habit 1: Ask Better Questions

Here's what I notice in most meetings: everyone's racing to have the smartest answer, but nobody's asking the right questions. People are often rewarded for having solutions without fully understanding a situation, and they don't necessarily exhibit the curiosity to probe more deeply. I see this all the time. Even in my own company I may get a call from a prospect saying they want to bring in a business acumen training course, but I'll pause and ask: "What's going on in your company, and what are some of your learning goals?" One question leads to another until I get to the heart of what someone's specifically looking for.

In your world, a customer may reach out to someone on your sales team with a request for a product, and the salesperson just provides the quote without understanding the underlying needs. Even in companies that teach consultative selling techniques, the salespeople forget that what customers ask for isn't always what they need. Strategic thinkers do a better job of asking "why" questions until they feel they can get to the root of the problem or situation, even if the process seems more tedious.

I learned this lesson the hard way early in my career. I was working as a product manager, and our biggest customer requested a specific feature enhancement. Instead of asking why they needed it or what problem they were trying to solve, I immediately started working with the engineering department to develop exactly what they asked for. Three months later, when we delivered the enhancement, the customer barely used it. It turned out that what they really needed was a completely different capability, but they had described their request in terms of the only solution they could imagine. If I had asked better questions upfront, we could have saved months of development time and delivered something truly valuable.

TRY THIS:
Here's how to put this into practice right away. Before an important meeting that you may be slated to lead, extend your thinking beyond the agenda and desired outcomes. Here's how you might prepare:

- Identify the key goals for the meeting. Is it just an update, solve a problem, or make a decision?
- To prepare, write down at least five probing (open-ended) questions that could reveal hidden issues, challenge assumptions, or clarify a problem. Avoid yes/no questions and try to shape questions that begin with why, how, or what if?
- For each question, anticipate a likely answer (knowing who's participating will help) and potential follow-up questions that probe more deeply.
- After the meeting, try to take some quiet time to reflect. Which questions offered you greater insight? What question sparked more discussion? Would you have asked different questions?

The power of better questions extends beyond individual conversations. Strategic thinkers use questions to challenge assumptions that entire organizations take for granted. They ask questions such as Why do we do it this way? What if our main competitor weren't a factor? and How would a new competitor approach this market? These questions can reveal blind spots that have been hiding in plain sight for years.

Asking better questions opens up new ways of seeing a situation. As you purposefully focus on questioning, you'll begin to recognize that your curiosity is contagious. Others will take note, and your credibility will build.

Those better questions that you're now asking are going to reveal patterns and connections you've never noticed before. That's our next habit.

Habit 2: Recognize Patterns and Make Connections

While most people see isolated events, strategic thinkers see patterns. Pattern mapping is the ability to connect observations across dissimilar areas or domains. Imagine how a store manager in a big retail chain might note that returns for a specific product are increasing. Some may just say that the product quality is bad, and the store chooses not to carry that product. Another might investigate the root cause for the returns and discover a connection between the returns and a change in packaging by a supplier's cost-cutting activity that caused more damages during shipping.

As you develop pattern recognition skills, you'll start spotting these connections across all areas of your business. You might notice that customer usage data shows declining engagement three weeks before churn typically occurs, or that product quality indicators correlate with seasonal supplier changes. You'll see how market segment shifts connect to changes in customer acquisition costs, or how employee satisfaction scores predict customer service ratings six months later. The key is training yourself to look for these non-obvious relationships across different data streams and time horizons. To make this more systematic, I've found that patterns typically emerge from three key domains.

1. *Market patterns* reveal how customers behave, what competitors do, and how entire industries evolve. When you understand these patterns, you're better equipped to identify future market opportunities before they become obvious to everyone else.

2. *Operational patterns* emerge from examining data across business processes, systems, and people. These patterns often reveal inefficiencies, bottlenecks, or opportunities that aren't visible when looking at individual metrics.
3. *Financial patterns* emerge from sales mix, pricing, margins, expenses, and profitability data. But the most insightful financial patterns often connect money flows to operational and market activities.

In my experience, the most valuable patterns emerge not from any single area, but from their intersections. When market shifts align with operational capabilities and financial realities, you've found strategic opportunity. When they misalign, you've identified strategic risk.

Cross-domain pattern discovery: Let me tell you about a software company I worked with that was losing sleep over customer churn. They were watching some of their biggest accounts, we're talking $10 million to $15 million per year clients, walk away after 12 to 24 months. On the surface, it looked like a product or service problem. I suggested we look at additional market data, but, more importantly, we reached out to these customers directly. (You'd think the account managers would have done this already, but apparently not.) The customers who churned weren't necessarily dissatisfied with the product. Nine of the companies were acquired. Interestingly, the acquiring firms already had competing software in place. This cross-domain pattern revealed that the real issue wasn't product quality or customer service, but other industry dynamics.

The discipline of non-obvious connections: Here's where most people go wrong with pattern recognition. They see something happening and immediately jump to the most obvious explanation. The sales team misses their quarterly target, and everyone assumes it's a pricing problem so there's a knee-jerk reaction to offer discounts. If customer complaints increase, someone blames product quality, so engineering is engaged to fix the problem, often taking their attention away from important projects. Strategic thinkers will pause and ask such questions as What else might explain this? What am I not seeing? and How might this connect to other things I've observed?

This discipline of looking for non-obvious connections often reveals the most valuable insights. In my experience, the first explanation that comes to mind is rarely the whole story. Sometimes it's not even the right story.

I once worked with a manufacturing company who was experiencing production slowdowns in the face of rising demand. In addition, employee turnover was increasing. The operations VP, without consulting the plant manager, believed the problem was low morale, and he decided to provide a 10 percent across-the-board raise. That didn't solve the problem. What we learned was that employees had ideas to improve efficiency and throughput, but their suggestions were being ignored by the plant manager. The employees thought that their opinions were not being heard, and they felt disrespected. The operations VP learned about this and ultimately brought the employees into the conversation. The problem wasn't money; it was about respect for the employees' expertise.

The key is developing what I call *pattern patience*. This is the ability to sit with observations without rushing to conclusions. Force yourself to consider at least three different explanations for any pattern you notice. Ask: If this obvious explanation is wrong, what else could be happening? Often the second or third possibility is where you'll find the real insight.

TRY THIS:
- Gather data, feedback, and observations related to a given topic. This could include revenue, customer comments, team insights, and market data.
- Look for recurring themes, anomalies, or associations. What keeps happening? What doesn't fit expectations?
- Draw pictures or map patterns that show connections between disparate observations or data that may offer clues.
- Reflect on what those patterns suggest about underlying causes or future risks for the organization. What questions come up? Are there assumptions that can be challenged?
- Share your insights with a colleague or members of your team and invite them to comment. Sometimes fresh eyes help complete the picture, and it encourages a collaborative environment.

By intentionally seeking and mapping patterns, you'll train yourself to move beyond surface-level understanding. You'll also develop a more holistic view of your challenges and potential opportunities.

Habit 3: Zoom In and Zoom Out

Here's what I see all the time: people get stuck either drowning in details or flying too high above them. Strategic thinkers fluidly move between both. They zoom in to understand specifics, then zoom out to see the bigger picture. To reinforce the point, zooming in means focusing on specifics, such as a series of data points, or drilling down on a process to assess root causes of a problem. Zooming out requires that you step back to view the situation from a more holistic standpoint. This technique allows you to look at connections within a function, across the organization, or how a situation might impact a company's strategy.

There is an art to doing this without getting stuck in either zooming in or zooming out. Too much time zooming in can lead to tunnel vision, where you'll miss trends or patterns. Too much time seeing the big picture can cause you to lose sight of an operational reality or a cultural nuance that needs real attention. In my research I've spoken with executives who are adept at moving effortlessly between these modes because they recognize that one is complementary to the other and can help them build greater levels of situational awareness. Strategic thinkers develop an almost instinctive sense of when to drill down and when to pull back.

During my product management career, I worked closely with engineers and product development people. Some of the most brilliant people were incredibly detailed, doing product design, programming, and developing products. However, they didn't always understand the case for the overall use of particular products, which sometimes caused products to be built with functionality that fulfilled an engineer's idea of something interesting but didn't necessarily solve the customer's problem.

While this orientation to detail is important, engineers can benefit from understanding the big picture of customer needs and problems (the zooming out mode), then zoom in to ensure that the product has just the right level of value-added functions.

TRY THIS:

- Before tackling a major challenge where an important decision is needed, try to identify what requires a detailed zoomed-in focus (e.g., key metrics, process details, stakeholder concerns). Write these down so they're visible and for you to assess them.

- Next, try to zoom out. Ask yourself: how does this issue connect to our broader goals? What external (e.g., market, competitor, customer) dynamics might influence what we're trying to do? What does success look like? What do we want to occur? Write these down, too.

- See if you can consciously switch back and forth between vantage points. You might need to zoom in again to check for overlooked details and then zoom out again to reassess until you feel you have a balanced viewpoint. Also, you can look at what you wrote down for visual clues (this is good practice!).

- As you work through an issues, set aside time in meetings or reviews to explicitly ask, "Are we too deep in the weeds? or Are we missing critical details by viewing things at a level that's too high?"

- Afterward, reflect on the situation. Did perspectives shift based on the assessment? Did your priorities change? Did decision options shift? Did you notice different things when you changed your vantage point?

When you can make this perspective shifting a habit, you'll develop the intuition to know when to dig deeper and when to step back, ensuring your decisions are informed, balanced, and aligned with both immediate needs and long-term vision.

Habit 4: Treat Every Situation as a Learning Opportunity

Here's what I've learned about most strategic thinkers: they're not the ones with all the answers. They're the ones asking all the questions. While most people rush to solve problems, strategic thinkers pause to understand them first. This resonates with me personally. Even as a kid, I always wanted to know how things worked and why. As my mom would say, my curiosity was insatiable. I frustrated her because I used to take things apart to see how they worked but didn't necessarily put them back together. I didn't realize this was a valuable characteristic until later in my life.

There are a couple of other areas where curiosity and learning go hand in hand. One has to do with asking for feedback. Most managers and leaders who interact with others, for example, on a project, will regularly check in with teammates or customers, or others who may offer fresh perspectives and guidance. As a lifelong learner, I have been grateful for any feedback I've received from asking questions, as well as feedback from people who cared, because I wanted to improve.

The other area is related to experimentation. Strategic thinkers who are also curious like to experiment. They know that not every idea will yield positive results, so they tend to try things out on a small scale to learn what works and what doesn't, and they adjust. In a world where business and market agility is prized, this mindset is beyond helpful.

Building Your Learning System

Let me share a practical framework I've developed for extracting lessons from every experience. When something doesn't go as planned, resist the immediate urge to fix it or find someone to blame. Instead, treat it as data that can improve your future decision-making. Start with these three questions:

1. What assumptions did we make that turned out to be wrong?
2. What signals did we miss that we should have caught?
3. What would we do differently if we faced this exact
   situation again?

Write down your answers. The act of documenting lessons makes them more likely to stick and become part of your strategic thinking toolkit.

Here's another technique I've found powerful: after any significant meeting, decision, or project milestone, spend 10 minutes doing what I call a "learning debrief." Ask yourself:

- What surprised me?
- What challenged my assumptions?
- What would I do differently next time? I

If you're leading others, invite them into this conversation. The goal isn't to critique performance; it's to capture insights while they're fresh.

Strategic thinkers also understand that learning isn't just about accumulating information. It's about changing your mind when new evidence emerges. They're comfortable admitting when they're wrong and adjusting their approach. This intellectual humility is a strength, not a weakness, because it allows you to course correct quickly instead of stubbornly pursuing failed strategies.

TRY THIS:
- Approach situations with three key questions: What don't I know yet? How can I learn more? Who can help me understand this better?
- After any significant meeting, decision, or project milestone, spend 10 minutes reflecting: What surprised me? What challenged my assumptions? What would I do differently next time? If you're coaching others, invite them into this conversation.
- Actively seek feedback from diverse sources. Don't just ask your usual work friends. Develop a network of people from other areas who might see things differently than you do. Ask specific questions: "Where do you think I'm missing something?" "What am I not considering?"
- Treat setbacks as valuable data, not failures. When something doesn't go as planned, resist the urge to immediately look for solutions. Instead, first ask: "What is this telling us about our assumptions, our market, our approach?"
- Start small experiments to test your ideas before committing significant resources. Create "safe-to-fail" trials where the cost of being wrong is low, but the learning potential is high.
- Keep a brief learning journal. Once a week, write down one thing you learned that changed your thinking about your business, your market, or your role.

By staying curious, asking for feedback, and testing ideas, strategic thinkers keep growing and, by association, help their teams to learn and grow.

Habit 5: Think Across Time Horizons

Here's what I see happening in most organizations: everyone's focused on this quarter's numbers while completely ignoring what those decisions mean for next year. Strategic thinkers think differently. They naturally balance immediate needs with long-term consequences. They understand that today's decisions create tomorrow's options or constraints. I'm not suggesting that you must be like a fortune teller or that you have to predict the future. It's about cultivating the discipline to ask If we do this today, what does it mean for us six months from now? Two years from now? What doors does this decision open or close?

Strategic thinkers also understand the concept of *temporal leverage*, or how timing multiplies impact. This reflects an idea that some decisions have a disproportionate impact over time. Hiring the right person today will impact your team's capability for years. Investing in a process improvement today can compound benefits over time. Building a relationship with a key customer today may open opportunities you can't even imagine yet.

The best strategic thinkers recognize that there are "temporal traps." This means that decisions that solve immediate problems or treat a symptom may create a more profound long-term issue. For example, if a company borrows money to fund a growth initiative with a fuzzy time horizon, short-term revenue may improve, but future financial flexibility could be constrained because of the financial obligation. Or a company will offer drastic discounts to win big deals and make the numbers for a quarter without understanding the long-term impact on profitability or even the company's premier reputation.

The key to developing this temporal discipline is to step back from immediate, seemingly urgent matters and consider the implications of your choices over various time horizons, even if this delay affects shareholder perceptions.

TRY THIS:

- Before making any significant decision, create a simple "time horizon analysis." Draw three columns labeled "Short-term (0–6 months)," "Medium-term (6 months–2 years)," and "Long-term (2+ years)."
- I strongly advise that you consider downstream implications of decisions to be made. My research shows that short-term urgent decisions don't consider long-term impacts, either because there's pressure to act (i.e., customer critical) or because there isn't enough data.
- For each option you're considering, write down the likely impacts in each time horizon. Focus on possible positive outcomes as well as potential negative impacts.
- Look for decisions that create positive momentum across multiple time horizons. This may help you build long-term value. It can be frustrating when long-term value is given away for the attainment of a short-term gain.
- Ask yourself: "What would I choose if I had to live with this decision for 5 years?" This single question often clarifies whether you're making a strategic choice or just reacting to immediate pressure.
- When facing quarterly or annual planning cycles, explicitly discuss time horizon trade-offs with your team.
- If you've mapped out the near-term and long-term implications of a decision, don't forget to audit or review the actual outcomes against agreed-upon plans. The best strategic thinkers continually assess progress, especially when they're not planning for that next quarterly business review.

When you develop the habit of thinking in terms of time horizons, you'll find yourself making decisions that compound positively over time, rather than creating recurring problems that consume your energy and resources.

MAKING STRATEGIC HABITS STICK (OR WHY THESE ARE HARD TO DEVELOP)

Building these strategic thinking habits isn't easy. If it were, everyone would be doing it. There are some very real obstacles that get in the way, and I want to acknowledge them because pretending they don't exist won't help you overcome them.

The Urgency Trap and How to Escape It

The biggest barrier I see is the relentless pressure to deliver immediate results. When your boss is asking for updates every day and your customers need answers now, it's hard to step back and ask deeper questions or look for patterns. The urgent consistently crowds out the important.

But here's what I've learned: the leaders who take time for strategic thinking create more time in the long run because they make better decisions that require fewer corrections later. One senior executive at a major bank said to me: "If you don't have time to consider the decision and its impacts now, when will you have time to fix it later?"

Your defense against the urgency trap is to start small and prove value quickly. Choose one habit and apply it to just one situation per week. When colleagues see that your strategic approach prevented a problem or revealed an opportunity they missed, they'll give you more space for this kind of thinking.

Organizational Cultures That Reward Speed Over Depth

Many companies inadvertently discourage strategic thinking by rewarding people who always seem to have the fastest answer or deliver results in the shortest time. Being the person who says, "I need to think about that" or "Let me ask a few more questions first" can feel risky when speed is valued over thoughtfulness.

If this describes your environment, here's how to navigate it: give the quick answer they want, then add value by going deeper. Say something like: "Here's what I think we should do immediately, and here's what I'd like to explore to make sure we're solving the right problem long-term." You're not slowing down the immediate response; you're adding strategic depth.

The Experience Paradox and Your Way Through It

Sometimes the more experienced you are, the harder it becomes to maintain a learning mindset. When you're seen as the expert, admitting you don't know something or asking for feedback can feel like you're showing weakness.

Strategic thinkers understand that confidence and curiosity aren't opposites, they're complementary. The most confident leaders are those secure enough to keep learning. When you frame questions as curiosity rather than ignorance, people see wisdom, not weakness. Say: "I'm curious about your perspective on this" rather than "I don't understand this."

YOUR IMPLEMENTATION STRATEGY

Knowing what to do and how to do it are two different things. Here's how to make these strategic thinking habits stick.

1. *Start with one habit.* Don't try to develop all five habits at once. Pick the one that feels most relevant to your current challenges and focus on it for 30 days or a time period that's attainable. Once it starts to feel more natural, add the next habit.

2. *Create environmental cues.* Some people (including me) like reminders. I use a sticky note to remind myself to think about a question before jumping into a discussion during a meeting, for instance. My note says: "Don't forget to question . . ." Here's another thing I do: I set 15-minute time blocks in my calendar. I usually write "thinking time" as the description. It gets me off the work treadmill so I can really think about things going on and perhaps consider some patterns of activity that might help me better understand situations that need careful consideration.

3. *Find a thinking partner.* Some leaders I spoke with have other people with whom they can share their thoughts about a situation. In my corporate leadership roles, I always had a few people who I could share ideas with who didn't run off and try to work on something because the boss talked about it. Sometimes I'd suggest we grab a drink or a snack after work. I also liked the morning coffee break, when I'd ask some

of my trusted colleagues to join me. Every executive I have interviewed has a close cadre of advisers or thinking partners they trust. Interactions with a thinking partner will help you improve your strategic thinking muscle.

4. *Track your progress.* At the end of each week, spend five minutes reflecting: When did you use these habits? When did you forget to use them? What triggered your strategic thinking? What got in the way? This reflection helps you recognize when strategic thinking is most needed and most difficult.

5. *Prove value early.* Choose low-stakes situations to practice these habits first. When you spot a pattern that helps avoid a problem or ask a question that reveals a hidden issue, document it. Build evidence that strategic thinking creates value, and you'll get more support for developing it further.

The goal isn't perfection; it's progress. Each time you apply one of these habits, you're strengthening your strategic thinking muscle and making it more natural for the next situation.

COMMON OBSTACLES AND HOW TO OVERCOME THEM

Even with the best intentions, you'll face obstacles as you try to develop these strategic thinking habits. Here are the most common ones I see and some practical ways to address them.

"I Don't Have Time for This"

This is the most frequent objection I hear. But strategic thinking doesn't require huge blocks of time. You can ask a better question in 30 seconds. You can look for a pattern while reviewing a report you're already reviewing. You can zoom out for two minutes before you make a decision. Start small and build from there.

"My Boss Wants Quick Answers"

If you work for someone who seems to value speed over depth, try this approach: give them the quick answer they want, then add, "And if we have a moment to dig deeper, I think there might be some additional considerations." Often you'll find they're grateful for the deeper thinking once they see its value.

"This Feels Too Theoretical"

If strategic thinking feels abstract to you, focus on the practical outcomes or results you want to achieve. Better questions lead to better information. Pattern recognition helps you predict problems before they happen. Zooming in and out helps you make more balanced decisions. A learning mindset helps you recover from mistakes faster. Keep connecting the habits to tangible results.

YOUR PATHWAY TO MORE IMPACTFUL STRATEGIC THINKING

When I first started developing these habits, I made the classic mistake of trying to do everything at once. This was frustrating. I'd try to ask better questions in one meeting or look for patterns while solving an urgent problem. The result? I felt scattered, and nothing really stuck.

My approach goes back to when I learned Transcendental Meditation and the power of focused practice. Whether you're learning to play a musical instrument or building physical fitness, strategic thinking tends to develop when you're deliberate, purposeful, and consistent in the mastery of one skill at a time.

When I started this journey many years ago, I created a plan and recorded it in a physical notebook. I numbered pages 1 to 30 so I could identify what I wanted to do and to record notes so I could track my progress. I also taught this approach to some of my direct-report employees. This systematic approach builds your strategic thinking muscle gradually and sustainably. Each week focuses on adding one habit. The system wasn't perfect because I found some overlap between habits, and real situations don't always fit neatly into weekly boxes. However, I think you'll greatly benefit from this self-education approach. It can serve you for the rest of your career.

The 5-Week Strategic Thinking Challenge

To help you get started, here's a practical 5-Week plan (shown in Figure 2.2) for building your strategic thinking muscle:

Figure 2.2 5-Week Strategic Thinking Challenge

Week 1 Better Questions	Week 2 Pattern Recognition	Week 3 Zooming Practice	Week 4 Learning Mindset	Week 5 Time Horizon Thinking
Focus on asking more thoughtful questions. Before each meeting, write down three probing questions. After each significant conversation, reflect on whether your questions led to deeper insights.	Start looking for patterns in your daily work. Set aside 15 minutes each Friday to review the week's data, feedback, and observations. What themes do you notice? What connections might others miss?	Before making any significant decision, consciously zoom in to examine the details, then zoom out to consider the broader context. Practice switching between these perspectives.	Focus on treating every situation as a learning opportunity. After each meeting or decision, ask yourself: What did I learn? What challenged my assumptions? What would I do differently?	Before making any significant decision this week, consciously consider the short-term, medium-term, and long-term implications. Practice temporal discipline by asking how each choice affects your options over time.

Week 6 and Beyond—Integration: Start combining all five habits. Use better questions to identify patterns across time horizons. Zoom in and out to validate your pattern recognition at different time scales. Approach everything with curiosity and a learning mindset while maintaining temporal discipline. Here's how you might do this:

- Month 1: Master your starting habits.
- Month 2: Expand to integrate multiple habits.
- Month 3: Apply integrated strategic thinking across multiple situations.

SUMMARY

The five habits of strategic thinking represent practical, learnable skills that transform how you approach complex challenges. Unlike abstract concepts, these habits can be developed through consistent practice and deliberate application to real business situations.

- *Asking better questions* shifts you from answer-provider to insight-generator. Instead of rushing to solutions, strategic thinkers probe deeper to understand root causes, challenge assumptions, and reveal hidden dynamics. This questioning discipline uncovers the real problems worth solving and prevents costly misalignment between what people ask for and what they need.

- *Pattern recognition* transforms scattered observations into strategic intelligence. By actively looking for connections across market, operational, and financial domains, you develop the ability to see trends and relationships that others miss. This capability allows you to anticipate challenges and opportunities before they become obvious to competitors.
- *Perspective flexibility (zooming in and out)* creates space for balanced decision-making. The ability to shift between detailed analysis and big-picture context ensures your decisions are both informed by specifics and aligned with broader objectives. This prevents tunnel vision while maintaining attention to critical details.
- *Learning mindset* accelerates your approach to strategic thinking excellence. When you view every situation with curiosity rather than certainty, you transform setbacks into valuable data and unexpected outcomes into learning opportunities. This intellectual humility enables faster course correction and continuous improvement of your strategic capabilities.
- *Time horizon thinking* prevents short-term optimization at long-term expense. By systematically considering immediate, medium-term, and long-term implications of decisions, you can make choices that create positive momentum across multiple time periods rather than solving today's problems while creating tomorrow's crises.

When practiced consistently, these habits reinforce each other and accelerate your strategic thinking development. Better questions reveal patterns across time horizons. Learning mindset improves your pattern recognition. Perspective flexibility enhances your temporal thinking. As each habit strengthens the others, your strategic thinking capability grows beyond what any single technique could achieve.

Implementation requires environmental support and gradual development. Strategic thinking habits can't be forced overnight but must be cultivated through consistent practice in real business situations. The key is recognizing opportunities that call for strategic thinking and developing the discipline to apply these habits even under pressure.

3

SYSTEMS THINKING: THE HIDDEN DISCIPLINE OF GREAT STRATEGISTS

Key Points

- When you see organizations as interconnected systems, you'll prevent expensive "fix one thing, break another" mistakes.
- Use the five systems lenses, and you'll find leverage points where small changes create big impacts.
- Think systemically, and you'll design solutions that strengthen the whole organization instead of optimizing isolated parts.

The Toyota style is not to create results by working hard. It is a system that says there is no limit to people's creativity. People don't go to Toyota to "work"; they go there to think.

— TAIICHI OHNO

The five habits from Chapter 2 build your strategic thinking foundation, but they address individual situations and decisions. Real business challenges rarely exist in isolation. Instead, they emerge from complex webs of relationships, feedback loops, and interdependencies that span departments, processes, and time horizons.

This is where systems thinking becomes essential. Systems thinking reveals how the pieces of your business connect, why solutions in one area often create problems in another, and where you can find leverage points that create positive change throughout your organization. Without systems awareness, even the best strategic thinking habits can lead to optimizing parts while accidentally breaking wholes.

WHEN GOOD INTENTIONS CREATE BAD OUTCOMES: THE REORGANIZATION THAT BROKE EVERYTHING

Mike ran a successful software company with 500 employees. Subscriber numbers and revenue had declined for two consecutive quarters, and the board was very concerned. I was brought in to assess what was happening, particularly with the core product team members from product management, development, marketing, and sales. My initial thrust was to understand market dynamics and evaluate whether there was a strategic issue.

What I discovered instead was that people were operating in silos with no clear accountability for results. Team members weren't aligned on common goals, and no one seemed accountable for business outcomes. The focus on product feature development was intense. However, when I asked what problems they were solving and for whom, they provided a host of inconsistent answers.

When I dug deeper, I learned that Mike and his senior team had undertaken a major reorganization nine months earlier. "We were trying to streamline operations, manage expenses, and demonstrate greater profitability," Mike explained. "The private equity firm was pressuring us to deliver better performance." He seemed certain the reorganization would have delivered the intended results.

I asked to see the before-and-after organization charts. Several changes stood out immediately: Product development now reported directly to Mike to save money by eliminating the role of head of development. They had also eliminated the chief marketing officer position, consolidating marketing, customer success, and sales under the chief operating officer. Mike explained the logic of bringing marketing and sales together for more effective coordination of efforts. But the results told a different story. Customer churn had doubled. New subscriber growth was 70 percent below plan. Employee satisfaction had plummeted, and several key players had left the company.

"Mike," I said at one point, "you assumed there was a straight line from where you were to where you wanted to be. Did anyone think about the ripple effects of these changes?"

The answer was clearly no.

The Systemic Consequences

Mike had fallen into the classic trap that derails even the smartest leaders: he treated a systems problem with linear thinking. Interestingly, the issues Mike faced resembled what experienced systems thinkers call *common systemic patterns*. These pop up all the time when leaders try to shift the burden of the problem (the reorganization) elsewhere, only to see the problem worsen. It's because they really don't understand the root cause of the problem.

Here's what the reorganization really created: Salespeople stopped collaborating with marketing because they said marketing didn't understand customers. Marketing didn't have the staff with the right level of expertise to do research on desirable customer segments. Moreover, they didn't have the resources and expertise to develop more sophisticated content, programs, and automation to generate leads. (The marketing budget was slashed during the restructure, eliminating the automated campaigns that generated 60 percent of qualified leads.)

The star product manager resigned because she no longer had a clear path to influence product direction. I caught up to her on LinkedIn to find out what happened, and she said that the developers were held in higher esteem and claimed to understand the product more than the product managers and that their every move was questioned by the technologists. Customer success teams, now merged with support, focused on reactive problem-solving instead of effective onboarding and proactive relationship building.

As it was explained to me by the leadership team, they thought that the reorganization made strategic sense. However, what I noticed beneath the surface was that Mike saw himself as the über-product manager and figured he had enough experience to guide the product, and the company, into the future. This to me was a flawed assumption. Also, the COO was consolidating power and influence, which perpetuated the company's "city of silos." What they did was to break the company because they didn't fully understand the connective tissue of the "system" of the business.

Mike learned the hard way what the best strategic thinkers know instinctively: in complex organizations, everything touches everything else. When you pull on one thread, the entire fabric can unravel. Linear thinking treats problems like isolated puzzle pieces. Systems thinking recognizes that those pieces are part of a living, breathing web of relationships.

That's the difference between leaders who create lasting change and those who unknowingly plant the seeds of their own problems. Mike is not alone. I see this pattern with many of my clients. By the way, a year later, the business imploded and was sold to another private equity firm.

THE SEDUCTIVE TRAP OF LINEAR THINKING

Let's be honest: linear thinking is seductive. It offers the illusion of control in a chaotic world. Problem A gets solution B. Step 1 leads to step 2. Cause leads directly to effect. It's clean, predictable, and satisfying.

So many processes are diagrammed as left-to-right with some branching logic thrown in for good measure. Linear thinking works beautifully in simple, stable environments. McDonald's perfected it for fast food. Assembly lines run on it. Even my morning coffee routine is fairly linear.

But most business challenges aren't like making coffee. They're complex, dynamic, and interconnected. Customers don't always follow what the product people record on the customer journey map. Competitors don't wait for your planning cycles, and the dynamism of markets don't wait to respond to your quarterly targets.

THE DEFAULT RESPONSE PATTERN

In my research and my work with clients, I find that, when faced with complexity, default responses tend to double down on linear solutions. Here are some that you may be familiar with.

- *Revenue is down:* Hire more salespeople or increase the marketing budget (sometimes it's fire the marketing people to reduce costs, and you know what that downward spiral looks like!).
- *Customer complaints are up:* Implement a quality improvement process.
- *Innovation is not providing creative solutions:* Change the development process.

Here's the problem: these solutions often create new problems. More salespeople without better leads will increase customer acquisition costs. Better processes without fixing root causes will create bureaucracy. Implementing a different development process without addressing the skills of key players and a culture of reactivity will create more internal focus and a race to the bottom.

I see this frequently in my work where experienced managers apply logical solutions that backfire spectacularly, not because the leaders lack intelligence, but because they're using the wrong thinking framework for the challenge at hand.

FUNDAMENTALS OF SYSTEMS THINKING

Systems thinking is the ability to see beyond isolated events and symptoms to understand the web of relationships and interdependencies that shape outcomes. It's recognizing that every part of a system connects to other parts, and that change rarely happens along straight lines.

Instead of "A causes B," systems thinking shows that A might influence B, which affects C, which loops back to influence A, creating an ongoing dynamic that's far more complex and far more powerful than any linear chain.

Think about your own organization for a moment. Your hiring decisions affect team culture, which influences productivity, which impacts customer experience, which drives referrals, which affects revenue, which determines your hiring budget, which circles back to influence future hiring decisions. That's a system at work.

The Principles of Emergence and Nonlinearity

Systems thinking also reveals two crucial principles missed by linear thinking: emergence and nonlinearity.

Emergence refers to systems where events emerge unexpectedly and alter patterns of behavior. Imagine if several geese veer off from their usual V formation and how that might alter their flight path. If this pattern becomes disruptive, the geese might have to alter their approach to yearly migration.

Nonlinearity refers to small changes that create disproportionate ripple effects in an organization. For example, let's say a company decides to switch to a hybrid working model to improve employee satisfaction. Employees love the arrangement, but no one thought about the building lease that runs for another ten years. Since a smaller space that would cost less money would be better, the company is stuck with the current lease with a lot of vacant space.

Both emergence and nonlinearity principles lead to disruptions that have negative consequences when leaders think linearly instead of systemically.

HOW SYSTEMS THINKING ENHANCES THE FIVE HABITS OF STRATEGIC THINKERS

Systems thinking naturally builds on the five habits of strategic thinkers, which were discussed in Chapter 2. Here's how:

Figure 3.1 How Systems Thinking Builds on the Five Habits of Strategic Thinkers

5 Habits	Applying Systems Thinking
Asking better questions	Reveals connections and feedback loops that all you to approach situations from different perspectives.
Recognizing patterns	Helps you to identify patterns across system boundaries (functions). It can also provide you with the wherewithal to assess market movements or shifts in customer preferences.
Zooming in – zooming out	Allows you to balance system details with interconnections and view things from a more holistic perspective
Learning mindset	If there's a failure, you'll learn more about system dynamics.
Looking across time horizons	This transforms how you view problems and provides you with insights about how systems truly work.

Systems thinking doesn't replace these habits. I believe it can super-charge them. It's the lens that transforms complexity into clarity and empowers you to design strategies for the real world.

FIVE ESSENTIAL LENSES FOR SYSTEMS THINKING

The best systems thinkers don't just stare harder at problems or situations; they view them through different lenses. (Sometimes, I use the term *multi-lens* to refer to the assortment of methods in which you can view various aspects of a system. I will use the term multi-lens in Chapter 6). Just as switching filters on a camera lens can reveal hidden details, these five lenses can contribute to your ability to see things that others might miss. Another way to look at this is by comparing mono-vision lenses from bifocals and trifocals. Each lens offers different ways of seeing things depending on what you're doing.

Lens 1: The Connection Lens

Most people see isolated problems. They may miss a deadline, a customer complaint, or perhaps a budget variance. *Strategic thinkers see the invisible threads connecting everything.* This lens reveals how a decision in marketing ripples through product development, affects customer service, and ultimately shows up in your financial results.

Beyond Direct Connections

Think about what happens when a company decides to speed up a product launch to beat a competitor to market. On the surface, it seems straightforward: faster time-to-market means competitive advantage. But through the connection lens, you see the web: rushed timelines stress the development team, which leads to more bugs or quality problems in the product, which increases support calls, which overwhelms customer service, which delays responses to paying customers, which hurts satisfaction scores, which impacts renewal rates six months later.

Most leaders focus on the obvious, direct connections: marketing spending affects lead generation, product development creates features, sales activities drive revenue. But the real insights come from the second- and third-order connections that are less obvious but often more powerful.

Mapping the Invisible Structure

The connection lens reveals that your organization's structure drives culture and behavior, but most of this impact is invisible until you map it out. When you trace how information flows, how decisions actually get made, and how work really gets done, you often discover that the informal connections matter more than the organization chart.

> **TRY THIS:**
> Pick a persistent challenge in your organization. Write it in the center of a whiteboard, then map every connection to it that you can think of: people, processes, systems, outcomes. Keep asking What else does this connect to? until you've identified at least 8 to 10 relationships. What surprises you about the web you've created?

Lens 2: The Feedback Lens

In my experience, most organizational problems are feedback loop problems in disguise. The feedback lens reveals the loops that drive system behavior, and understanding them is crucial for any leader who wants to create sustainable change.

Understanding the Two Types of Loops

There are two types of loops that shape every organization. *Reinforcing loops* amplify change where an action creates conditions that intensify the original action. Success breeds success, but failure breeds failure too. A great product gets positive reviews, driving more sales, which provides more funding for development, which improves the product further. But the same dynamic works in reverse: quality problems lead to negative reviews, reduced sales, budget cuts, and further quality problems.

Balancing loops resist change. When a talented employee becomes overloaded with work, performance eventually drops, frustration builds, and either the workload gets redistributed or the employee leaves. The system naturally pushes back against the overload, but sometimes not before real damage is done.

Beyond Reactive Feedback

Here's what separates good systems thinkers from great ones: they don't just react to feedback loops, they design them. Think of a restaurant that uses today's customer preferences, inventory patterns, and staff observations to inform tomorrow's prep decisions, staffing levels, and menu choices. They're not just learning from what happened; they're anticipating what's coming next.

The best organizations build both feedback loops, which help them learn from results, and feedforward mechanisms, which help them anticipate and prepare for what's ahead.

> **TRY THIS:**
> Think about a recurring problem in your organization. Draw a simple loop: What leads to the problem? What does the problem lead to? Keep following the chain until it circles back to the original issue. Is this loop reinforcing the problem or trying to balance it?

Lens 3: The Time Lens

This might be the most important lens for leaders, yet it's the one I see misapplied most often. The time lens reveals how delays distort cause and effect, and why so many good strategies get abandoned too soon, while bad strategies last too long.

The Delay Problem

In systems, outcomes often lag behind decisions by weeks, months, or longer. A cost-cutting decision might improve margins immediately but damage customer relationships that won't show up in churn data for six months. An investment in employee development may not impact productivity for a year. A marketing campaign can look like a failure for months before the delayed positive effects kick in.

Without understanding these delays, leaders make two critical errors: they overreact to short-term noise and abandon good strategies before they have time to work. I've seen CEOs kill promising initiatives after just a few months because they expected immediate results. I've also seen them double down on failing strategies because early indicators looked positive.

The Quarterly Pressure Trap

Many people who work in publicly traded companies are familiar with this and often frustrated by what they see as "making the numbers" for the quarter, which starves the company of investments that could build future capabilities. You need to understand the natural rhythm and response times of your system before you can determine if your interventions are working.

> **TRY THIS:**
> Think about a recent decision your organization made. Create a timeline: What was supposed to happen immediately? What might not show up for three to six months? What consequences may not be visible for a year or more? Are you measuring success on the right timeline?

Lens 4: The Assumption Lens

When systems aren't working, the problem is rarely poor execution. It's usually a flawed mental model driving the execution. The assumption lens helps you see the often-invisible beliefs that shape how systems work.

In my career and in my consulting work, I've seen this pattern repeatedly in business cases and strategic plans. Teams will present what looks like rigorous analysis, but when I dig deeper, I discover that crucial data is missing, or key relationships are based on educated guesses rather than facts. This isn't necessarily bad. You see, in business, not everything can be reduced to a number, and you often must make decisions with incomplete information.

Like the paleontologists I mentioned earlier who use clay to fill gaps in dinosaur skeletons, you need to create realistic hypotheses to bridge missing pieces of your business analysis. The key is being explicit about what you're assuming versus what you know for certain. Make sure you document your assumptions as well. I've watched too many strategic initiatives fail because teams forgot which parts of their analysis were clay and which were actual bone.

Uncovering Hidden Beliefs

Every organization operates on mental models about people, performance, competition, and value. These models determine policies, processes, and decisions. They're so embedded in how we work that we rarely question them until something breaks.

Consider a manufacturing company struggling with product quality issues. The leadership team's initial reaction may be to blame employees, assuming that "our people don't care about quality." But when you dig deeper, you often find that the real problem is baked into the system itself: the operations team is rewarded for speed, so inspections get cut. Meanwhile, the supply chain team sources cheaper components to hit cost targets, which also hurts quality.

The hidden assumption wasn't that people didn't care, it was that speed and cost mattered more than quality. That belief was built right into their rewards and processes.

Systems Built on Outdated Assumptions

The assumption lens helps you spot the underlying beliefs so you can change them. Many organizational frustrations stem from systems designed around assumptions that may have been true once but no longer apply.

> **TRY THIS:**
> Pick a policy or process in your organization that frustrates people. Ask: What belief about people or performance is this based on? Then ask Is that belief really true? Often you'll find systems built on outdated or incorrect assumptions.

Lens 5: The Leverage Lens

The idea of leverage points comes from systems thinkers who spent years looking for ways to make changes stick in complex systems. They learned that not all intervention points are created equal since some changes have a greater impact than others.

Finding the Right Lever

Think about leverage like moving a boulder. You could try to push the boulder with all your strength and not succeed in moving it. Or you could use the right tool, a lever, and place it at the right leverage point to increase the force needed to move the boulder. When leaders get stuck focusing on the obvious, as in setting new targets or budgets, for example, they're often missing the real opportunity: changing the rules, goals, or beliefs that hold the system in place.

In my experience, the highest leverage usually comes from shifting what the system is actually trying to achieve or questioning the beliefs that define what success looks like.

A Practical Example

Let's say you're leading a logistics company, and your data show that order fulfillment times are too slow. The easy, low-leverage fix might be to hire more packers or add another shift. But what if the real bottleneck is buried in the process? Maybe shipments are delayed because approval for high-value orders requires multiple sign-offs, adding hours to every transaction. A small change such as streamlining approvals or automating a simple risk check could remove that bottleneck and speed up the entire system.

TRY THIS:
Think of a persistent problem in your organization. Instead of asking What can we add or fix? Ask: What small change in approvals, information flow, or decision-making could remove the biggest bottleneck? Look for leverage points where a minor tweak could create a major shift.

Understanding systems through these five lenses is powerful, but insights without action don't solve problems. Each lens reveals something important about how your organization really works. The five tools that follow will help you translate those insights into interventions that actually stick.

FIVE PRACTICAL TOOLS FOR SYSTEMS THINKING

Tool 1: The Connection Web

What it does: Maps how different people, teams, processes, and outcomes connect so you can spot hidden bottlenecks and unexpected leverage points.

How it works:

1. Write your biggest challenge in the center of your workspace (paper, whiteboard, or digital tool like Miro or Lucidchart, etc.)
2. Around it, write down all the people, teams, processes, outcomes, constraints, and resources that might be related to this challenge
3. Draw lines connecting these factors to your central challenge
4. Draw additional lines connecting factors to each other where relationships exist
5. Look for patterns: What has the most connections? Where do connection clusters form? What seems isolated but shouldn't be?

Quick example: A team discovered their communication problem was actually a decision-rights problem. Nobody knew who could approve changes, so everything stalled.

When to use this tool: Use the connection web when you're facing a recurring problem that seems to have multiple causes, when you suspect a problem is more complex than it appears, or when solutions in one area keep creating problems elsewhere. It's especially valuable before major organizational changes or when trying to understand why good initiatives aren't getting traction.

> **TRY THIS:**
> Pick your most frustrating recurring issue. Map every connection you can think of, then ask others on your team to add connections you missed. The surprising connections often reveal the real leverage points.

Tool 2: The Problem Iceberg

What it does: Reveals what's really driving problems by looking beneath surface symptoms to find deeper patterns and root causes through four distinct levels of analysis.

How it works: Work through these four levels systematically:

> **Level 1—Events:** What happened? (the visible symptoms)
> **Level 2—Patterns:** What trends do you notice? What keeps repeating?
> **Level 3—Structures:** What rules, processes, or systems create those patterns?
> **Level 4—Mental Models:** What beliefs, assumptions, or mind-sets drive those structures?

Start with the obvious problem (Level 1) and keep asking "What's causing this?" until you reach the underlying beliefs (Level 4).

Quick example: High turnover looked like a compensation problem (Level 1), but the real issue was a promotion system that ignored internal talent (Level 3), driven by a belief that external hires bring fresh thinking (Level 4).

When to use this tool: Apply the problem iceberg when you keep solving the same problem over and over, when quick fixes aren't working, or when you suspect you're treating symptoms rather than root causes. It's most powerful for persistent organizational issues like turnover, quality problems, or communication breakdowns.

TRY THIS:
Take any recurring problem and work through all four levels. The fourth level, mental models, is usually where you find the real leverage point for lasting change.

Tool 3: The Loop Tracker

What it does: Identifies feedback loops that either make problems worse (reinforcing loops) or try to maintain balance (balancing loops), so you can design solutions that work with system dynamics instead of against them.

How it works:

1. Start with a recurring problem
2. Ask: "What causes this problem to happen?"
3. Then ask: "What happens as a result of this problem?"
4. Keep following the chain of cause and effect
5. Continue until the effects loop back to influence the original problem
6. Identify if it's a reinforcing loop (making the problem worse) or balancing loop (trying to correct the problem)

Reinforcing Loop Example: Pressure to hire quickly → weak screening → poor cultural fit → increased turnover → more pressure to hire quickly (the loop amplifies the problem)

Balancing Loop Example: Sales drop → management cuts prices → short-term sales increase → lower margins → pressure to raise prices → sales drop (the system tries to self-correct)

Quick example: A startup's rapid hiring created a negative reinforcing loop that accelerated turnover until they broke it by improving their screening process.

When to use this tool: Use loop tracking when you notice problems that seem to get worse despite your efforts to fix them, when solutions create unintended consequences, or when you want to understand why certain patterns keep repeating in your organization.

TRY THIS:

Map one problem that keeps coming back. Trace the complete loop. If it's reinforcing (making things worse), find where to break the cycle. If it's balancing (self-correcting), work with it rather than fighting it.

Tool 4: The Timeline Reality Check

What it does: Prevents you from killing good strategies too early or sticking with bad ones too long by mapping realistic timelines between actions and results, including early warning indicators.

How it works:

1. For any initiative, create three timeline buckets:
   * Immediate (1–30 days): What should you see right away?
   * Short-term (1–6 months): When should early results appear?
   * Long-term (6+ months): When should full results be evident?
2. For each time period, identify:
   * Leading indicators (early signals that predict success)
   * Lagging indicators (actual results)
   * Warning signs (signals that something's wrong)
3. Build in checkpoints at each stage to assess progress without panic

Quick example: A customer loyalty program might look like a failure after three months, but the company realized it takes six months for customer behavior changes to show up in sales data. Their early leading indicators (program enrollment and engagement) were positive.

When to use this tool: Apply timeline reality checks before launching any strategic initiative, when stakeholders have unrealistic expectations about results, or when you're tempted to abandon a strategy because you're not seeing immediate results.

> **TRY THIS:**
> For your next big decision, create a timeline showing when you'll realistically see results. What early indicators will tell you if you're on track? Set calendar reminders for each checkpoint to resist premature judgment calls.

Tool 5: The Change Multiplier

What it does: Finds the highest-impact intervention points where small changes create big system shifts across multiple levels.

How it works: Map potential changes across different levels: numbers/metrics, resources, rules/policies, goals, and core beliefs. Look for changes that would influence multiple levels.

Quick example: A consumer products company cut costs that had a negative impact on product quality. They also reduced the staff of the customer service department. The result was a decrease in customer satisfaction, increased product returns, and reduced revenue. The diagram shown as Figure 3.2 shows the entire picture, from the trigger point, the solutions, and the multiplier over six months.

Figure 3.2 The Change Multiplier

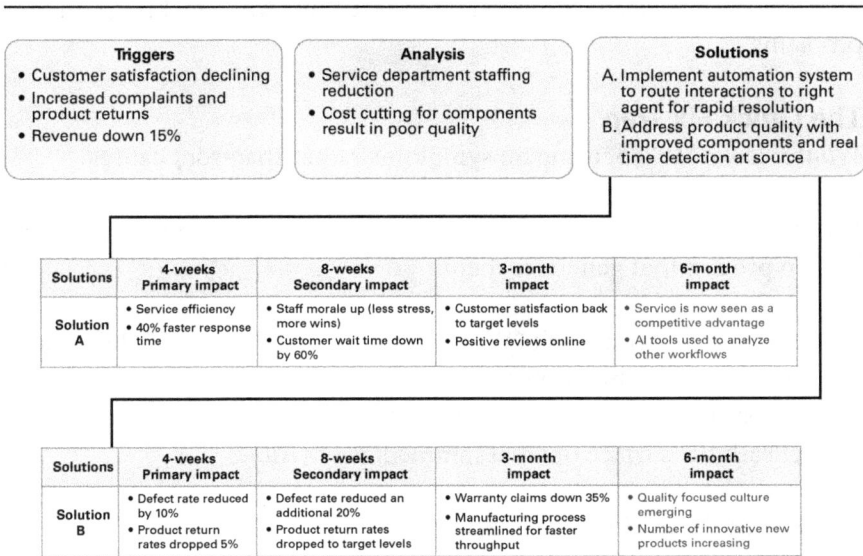

Triggers	Analysis	Solutions
• Customer satisfaction declining • Increased complaints and product returns • Revenue down 15%	• Service department staffing reduction • Cost cutting for components result in poor quality	A. Implement automation system to route interactions to right agent for rapid resolution B. Address product quality with improved components and real time detection at source

Solutions	4-weeks Primary impact	8-weeks Secondary impact	3-month impact	6-month impact
Solution A	• Service efficiency • 40% faster response time	• Staff morale up (less stress, more wins) • Customer wait time down by 60%	• Customer satisfaction back to target levels • Positive reviews online	• Service is now seen as a competitive advantage • AI tools used to analyze other workflows

Solutions	4-weeks Primary impact	8-weeks Secondary impact	3-month impact	6-month impact
Solution B	• Defect rate reduced by 10% • Product return rates dropped 5%	• Defect rate reduced an additional 20% • Product return rates dropped to target levels	• Warranty claims down 35% • Manufacturing process streamlined for faster throughput	• Quality focused culture emerging • Number of innovative new products increasing

When to use this tool: Use the change multiplier when you have limited resources but need significant impact, when you're looking for the highest-leverage intervention points, or when you want to ensure your changes create positive momentum across multiple areas.

TRY THIS:
For your biggest challenge, brainstorm solutions at each level. Which small change could create the biggest ripple effect across your organization?

These five tools give you practical ways to apply systems thinking, but even experienced leaders can fall into predictable traps when tackling complex problems. Understanding these tools is one thing. Knowing where people typically go wrong is another. Let me show you the three most common mistakes that derail even well-intentioned systems thinking efforts.

AVOIDING COMMON SYSTEMS TRAPS

Even experienced leaders can fall into these traps when tackling complex problems.

The Quick Fix Trap
What it looks like: Focusing on symptoms rather than root causes.

Example: Adding more staff to handle complaints instead of fixing the broken process that generates them.

How to avoid it: Always ask "What's creating this problem?" Before asking how do we fix this problem? Use the problem iceberg to go deeper than the surface symptoms. Like an actual iceberg, 90% of the problem lies beneath the surface or what's immediately visible.

The Local Optimization Trap
What it looks like: Improving one part of the system while accidentally hurting another.

Example: Reducing costs in one department but increasing workload elsewhere, creating bigger problems downstream.

How to avoid it: Use the connection web (a visual tool you draw to map relationships between people, processes, and outcomes) to trace connections before making changes. Ask "If we optimize this area, what happens everywhere else?" Map the ripple effects of your proposed change across all connected elements before implementing. As I mentioned earlier, this meshes nicely with the leverage lens. Once you see all the connections in your web, you can identify which ones offer the highest impact intervention points rather than creating unintended consequences elsewhere in the system.

The Delay Misreading Trap

What it looks like: Giving up on a good strategy too soon or sticking with a bad one too long because you don't see immediate results or you're seeing temporary positive results from a fundamentally flawed approach.

How to avoid it: Use delay mapping to set realistic expectations for when results should appear. Create leading indicators (early signals) that predict future outcomes before lagging indicators show up. Build in regular checkpoints that account for system delays. For example, if a new sales process takes three months to show revenue impact, measure pipeline velocity and conversion rates at weeks 2, 4, and 8. Make sure to communicate these time delays to stakeholders so they don't abandon the strategy too early. Most importantly, resist the pressure to declare failure or pivot before your delay map says you should see results. Also, document your expected delay periods upfront and stick to them unless leading indicators suggest fundamental problems.

Avoiding these traps is crucial because systems thinking isn't just an analytical exercise—it's the foundation for everything that follows. In the next chapter, we'll explore how the right mindset transforms systems thinking from a useful framework into a powerful competitive advantage.

SYSTEMS THINKING IN ACTION: THREE BUSINESS SCENARIOS

To help you see how systems thinking transforms problem-solving in practice, let me share three scenarios that demonstrate the difference between linear and systems approaches. In each case, the obvious solution would have worsened the underlying problem, while systems thinking revealed intervention points that addressed root causes. These examples show how the five lenses and tools work together to uncover insights that traditional problem-solving approaches miss entirely.

Scenario 1: The Marketing Budget Dilemma A software company's marketing director requests a 40% budget increase because lead generation is down 25%. Linear thinking says: "More budget = more leads." Systems thinking reveals that recent product updates confused the messaging, sales teams changed their qualifying criteria, and the customer success team is too busy to nurture leads properly. The real solution isn't more marketing budget, it's aligning the entire customer acquisition system.

Scenario 2: The Innovation Stalemate A product team consistently misses product launch deadlines despite having talented developers, marketers, and product people. Linear thinking says: "Add more engineers or extend deadlines." Systems thinking uncovers that requirements keep changing because the product managers don't understand customer needs, which happens because the customer success team rarely shares insights, which occurs because they're measured on support tickets resolved, not customer insights gathered. The leverage point is changing how customer success is measured and rewarded.

Scenario 3: The Cost-Cutting Cascade A manufacturing company cuts maintenance staff to reduce costs. Six months later, equipment downtime increases, production delays rise, and customer complaints spike. Linear thinking says: "We cut too much; hire back some maintenance staff." Systems thinking reveals that the cuts broke the feedback loop between maintenance, production, and quality control. The solution requires not just restoring staff, but redesigning how information flows between these functions.

Each scenario shows how systems thinking reveals intervention points that linear thinking misses entirely.

SUMMARY

After working with hundreds of leaders who've struggled with complex business challenges, I've seen how transformative it can be when you view systems instead of just events. The connection lens has saved my clients from countless expensive mistakes; those painful situations where you fix one department's problem only to create bigger problems elsewhere.

Understanding feedback loops changed how I approach every business challenge. When you can spot the reinforcing loops that make successful companies stronger and failing ones weaker, you stop fighting against system dynamics and start working with them. The balancing loops teach you when to persist and when your system is naturally pushing back for good reasons.

I've watched too many good strategies fail because leaders didn't understand the time lens. Most strategic failures aren't execution failures, they're timing failures. Leaders abandon good strategies too early or stick with bad ones too long because they don't account for system delays.

The assumption lens might be my favorite because it reveals the invisible beliefs driving organizational behavior. In my consulting work, I've found that when systems aren't working, it's rarely an execution problem, it's usually outdated or faulty assumptions that are built right into how people think and work.

The leverage lens is where systems thinking becomes powerful. Instead of throwing resources at symptoms, you find those intervention points where small changes create big impacts across your entire organization.

The five tools give you practical ways to apply these insights, but here's what I've learned: knowing about systems thinking and applying it when your hair's on fire are two different things. That's why your mindset matters so much. It's the foundation that makes everything else work when it matters most.

Now it's time to develop the mental foundation that makes everything else work when it matters most.

CHAPTER

4

MINDSET MATTERS: THE MENTAL MODELS THAT MAKE OR BREAK STRATEGIC THINKING

Key Points

- The leaders who say "I don't know yet" often make better decisions than those who claim to have all the answers.
- Best practices from other companies can destroy your competitive advantage if you apply them without the proper context.
- The more you try to control business outcomes directly, the less influence you actually have.

To the man with only a hammer, every problem looks like a nail.

—CHARLIE MUNGER

If you're reading this book and you've come this far, you're probably someone who's been told to think more strategically. Whether you're an ambitious manager or a leader who's interested in a more purposeful approach to your career development, mindset matters.

But what does it take to have a strategic mindset? How can you develop this mindset, especially with what I've discussed in the first three chapters of this book? Moreover, how do you take a break from running on the treadmill of your job when you're putting out fires, attending meetings, and trying to keep your head above water?

I've been there. I've sat in those meetings where the word *strategy* is tossed around like confetti, but everyone is weighed down by an endless barrage of problems that require immediate attention. Who has time to think?

Here's the truth: strategic thinking isn't just a skill or a host of techniques. Like a theater production, you need a stage or platform on which to stand. In strategic thinking, I believe this platform is your mindset. Your mindset allows you to deal with complexity, ambiguity, and change. It also supports your ability to connect the dots and make choices that matter in the long term, not just the next quarter.

Think about it: you can master pattern recognition (Chapter 2), but if you approach every pattern with the assumption that you already know what it means, you'll miss crucial insights. You can apply the five lenses (Chapter 3), but if you're not open to what those lenses reveal, you'll see only what confirms your existing beliefs. You can ask better questions, but if you're not prepared to hear uncomfortable answers, those questions become exercises in confirmation bias (seeking information that supports what you already believe).

A strategic mindset is the foundation that makes everything else work when it matters most.

A PERSONAL LEARNING MOMENT

In all honesty, I'm a work in progress. In the early part of my career, I thought I had to have a lot of answers. Sometimes I'd be too active in meetings, until one day my manager suggested that I say nothing in meetings, just for a while. He said to me, and I'll paraphrase: "You need to sit back and let everyone talk, then assess the situation before deciding what to contribute." He suggested that I take notes and write down my thoughts on paper without saying anything. That was a major revelation for me and was great advice.

When I first started digging into strategic thinking, I realized that the best leaders weren't always the ones with an immediate answer or the smartest ones in the room. They were the ones who would pause, step back, and ask, "What's really going on here? Is there anything I'm missing? How could we better address this issue?" Simply put, they had an uncanny way of making sense of situations based on what they knew, on what they had learned, and on their experience. They had built a tool kit of mental models.

This experience taught me something crucial: strategic thinking isn't about having all the answers. To me, it's about having the ability to build mental models that can help you find better answers. It's about developing what Charlie Munger, the late vice chairman of Berkshire Hathaway, called a "latticework of mental models."

MENTAL MODELS: YOUR SECRET WEAPON

If you're not familiar with the term, think of mental models as lenses through which you view and make sense of the world around you. We all have them, whether we realize it or not. The key is to become aware of which ones you're calling on in a situation, and to be alert to the formation or reformation of these models as you expand your base of experience.

I've been a fan of the conglomerate Berkshire Hathaway for decades. I eagerly await its annual conference in Omaha and have loved the things that its chairman, Warren Buffett, and (the late) Charlie Munger had to say. If you're not familiar with the company and these stellar business minds, I urge you to learn about them.

Munger was Buffett's partner, and I'm going to present what I cobbled together from a series of quotes from Charlie that I show in my business acumen training course:

> You can't really know anything if you just remember isolated facts and try to bang 'em back. If the facts don't hang together on a lattice-work . . . you don't have them in usable form . . . you've got to have models in your head. And you've got to array your experience both vicarious and direct on this latticework of models. . . . And the models have to come from multiple disciplines because all the wisdom of the world is not to be found in one little academic department. . . .

Building your own latticework requires that you are curious and open, that you're willing to study, ask questions, learn how things work, and challenge your own assumptions about the world. It's about building relationships with people in different departments to understand how they do what they do, and why. It's about understanding how your customers do what they do, what problems they're trying to solve, and perhaps how your company's products solve those problems. Your latticework is what you make it, and you'll be surprised how it can grow if you're aware that it's a vital part of your strategic thinking journey.

WHEN SMART LEADERS HIT MENTAL WALLS

Let me share a story that illustrates why mental models matter so much. I was working with a CEO named David, who had built his reputation on being decisive. Fast decisions, clear direction, unwavering confidence: his team loved it . . . until they didn't.

The company was losing ground to a smaller competitor, and David's response was typical: "We need to execute better. Work harder. Focus more." When I suggested the market might be shifting in ways that required a different approach, he cut me off. "I've been doing this for twenty years. I know what works." (Yes, there are still people who say this kind of stuff.)

Six months later, they lost their biggest customer to that same smaller competitor. I was asked to meet with the customer to assess what happened. I'll paraphrase what he said to me: "Those guys kept trying to sell us the same stuff and didn't understand that we shifted our strategy and business model. They didn't notice because they stopped coming around and they lost track of our business."

David's problem wasn't his intelligence or his experience. It was his mental models. He had developed his own latticework that caused him to equate leadership with immediate answers, and he stuck to a pattern of thinking he'd developed. It got stale. When the world changed around him, his mental models became a prison that prevented him from seeing new possibilities.

This story illustrates what I call "mindset lock." It occurs when successful mental models become rigid assumptions that prevent adaptation. David's decisive leadership style had worked for years, creating a feedback loop that reinforced his belief that quick decisions and unwavering confidence were always right. But mental models that work in stable environments often fail in dynamic ones.

The strategic thinking habits from Chapter 2 could have helped David recognize patterns of market change. Systems thinking from Chapter 3 could have revealed how shifts in client needs connected to broader industry trends. But his mindset prevented him from applying these tools effectively. He wasn't curious about new patterns because he believed he already understood his market. He didn't explore system connections because he was focused on controlling outcomes through force of will.

The most effective and impactful leaders I know aren't the ones who are always right. They're the ones who can change their mental models when the facts change. They understand that in a world where change is the only constant, rigid thinking becomes a liability.

THREE FUNDAMENTAL MINDSET SHIFTS

After working with hundreds of leaders, I've identified three fundamental shifts in mental models that separate strategic thinkers from everyone else. These aren't personality traits or natural talents. They're learnable tools that anyone can develop with intention and practice. Think of these as part of a mindset bridge, as illustrated in Figure 4.1.

Figure 4.1 Mindset Bridge Characterizing Three Fundamental Shifts

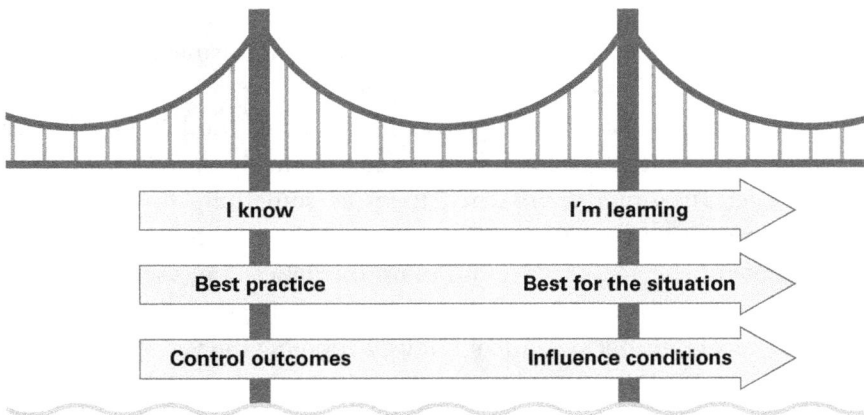

I know	I'm learning
Best practice	Best for the situation
Control outcomes	Influence conditions

Shift 1: From "I Know" to "I'm Learning"

Most leaders feel enormous pressure to have the answer. They're expected to be the expert, the one with solutions, the person who provides certainty in uncertain times. This creates what I call the "knowing trap." This is the belief that leadership means always having the right answer.

Strategic leaders flip this completely. They see uncertainty not as a threat to their credibility but as information to be explored. They're comfortable saying "I need to understand this better" and "What am I missing here?"

This shift transforms how you apply every strategic thinking tool. When you're asking better questions, the learning mindset makes you genuinely curious about the answers rather than seeking confirmation of what you already believe. When you're applying the five lenses from Chapter 3, you're open to seeing what each lens reveals rather than filtering insights through existing assumptions.

I saw this difference clearly with two division heads at the same company, both facing high employee turnover. The first leader immediately diagnosed the problem: "People want more money." She got approval for raises. Turnover dropped temporarily but returned within months, now with higher payroll costs. She didn't know that money is not a motivator.

The second leader said, "I don't know why people are leaving, and that bothers me. Let me find out." He spent weeks talking to employees and discovered the real issues: poor manager–employee relationships and unclear career paths. His solution took longer but solved the problem permanently with the help of a knowledgeable HR business partner and his chief operating officer.

The difference wasn't intelligence. It was mental models. One leader approached problems as something to solve quickly based on existing knowledge. The other approached them as something to understand deeply, even if that meant admitting ignorance first. If you recognize this in your own behavior, try to zoom in on the details, but then zoom out and try to assess the situation from a different vantage point.

This second leader was unconsciously applying the variance analysis approach, which I'll discuss in Chapter 5. Instead of accepting turnover as a given, he treated solving the problem as detective work. He was looking for patterns in who was leaving, when they left, and what they said about their reasons. His learning mindset enabled him to see turnover not as a simple compensation problem, but as a symptom of deeper systemic issues.

This shift requires overcoming the fear that admitting uncertainty will damage your credibility. In reality, leaders who demonstrate curiosity and learning orientation build more trust because their eventual decisions are better informed and more thoughtful.

The learning mindset also changes how you handle failure and unexpected outcomes. Instead of defending decisions that didn't work, you treat them as data that improves future decision-making. This creates what psychologist Carol Dweck calls a "growth mindset." This is the belief that abilities and intelligence can be developed through learning and practice.

TRY THIS:
The next time someone brings you a problem, resist giving advice for the first five minutes. Instead, ask questions to understand what's really happening. Notice how this changes both the conversation and the quality of your eventual solution.

In strategic thinking terms, the learning mindset enables you to update your mental models based on new information rather than defending outdated assumptions. This is crucial because business environments change constantly, and what worked yesterday may not work tomorrow.

Shift 2: From "Best Practice" to "Best for the Situation"

The business world is obsessed with best practices. In my diagnostic work with clients, I work with them to understand how specific business practices impact outcomes. Each company is different, and there are different levels of practice maturity, or the degree to which a practice is consistently applied. What I learned is that strategic leaders question one-size-fits-all best practices and seek a deeper context before finding a solution.

This shift connects directly to the systems thinking from Chapter 3. When you understand that every organization is a unique system with its own culture, constraints, capabilities, and market position, you realize that solutions must be adapted to fit specific system conditions.

I experienced this firsthand with two companies that both needed to revamp their strategic planning process. The first was an established technology company with experienced executives, a stable market position, and predictable revenue streams. I was part of a team that assessed their current strategy formulation process. The team was able to fine-tune the process with enhanced market analysis, detailed financial projections, and a better product portfolio allocation protocol. Ultimately, the assessment resulted in a greater number of initiatives that fulfilled the agreed-upon strategic intent.

The second company was a fast-growing startup in a rapidly evolving market with limited historical data and constantly shifting competitive dynamics. The company leaders tried to implement a traditional strategic planning process, but they were failing. We were asked to do a review of the process and learned that an annual protocol was not practical for business where the tides of the market shifted without notice. We were able to help them create a streamlined 90-day goal-setting process that we called "strategic sprints" with mini-business cases to justify and validate near-term investments that allowed them to quickly pull the plug on some projects that lost strategic value and shift resources more rapidly to projects that made more strategic sense. The idea was to build strategic thinking mindsets instead of maintaining the rigorous process that crippled their ability to compete.

The startup's solution demonstrates contextual thinking in action. Instead of forcing their organization into a planning framework designed for stable environments, they designed a process that fit their system's characteristics: speed, uncertainty, limited resources, and the need for rapid learning.

Strategic thinking leaders don't ask, "What's the best practice?" They ask, "What's the best practice for our specific situation, with our specific people, facing our specific challenges, at this specific time?"

This requires developing what I call "situational intelligence." This relates to the ability to read the unique characteristics of your environment and adapt solutions accordingly. It means understanding your organization's culture, constraints, capabilities, and competitive position well enough to modify approaches that work elsewhere.

Contextual thinking also applies to timing. A solution that works in a growth market might fail in a recession. An approach that succeeds with experienced teams might overwhelm new employees. Strategic thinkers consider not just what to do, but when and how to do it given their specific circumstances.

> **TRY THIS:**
> Before implementing any solution from another company or department, spend 15 minutes listing what makes your situation unique. How might these differences require you to modify the approach?

This shift requires developing comfort with ambiguity and the discipline to analyze context before applying solutions. It means being willing to modify or even abandon approaches that work elsewhere if they don't fit your situation.

Shift 3: From "Control Outcomes" to "Influence Conditions"

This might be the hardest shift because it challenges a fundamental assumption about leadership. Why? Because it's assumed that senior leaders are supposed to produce intended results. Unfortunately, most senior leaders don't get involved in enough details, so they suffer when an outcome isn't fulfilled and celebrate when they hit the mark. While they can't control everything, they can influence the conditions that improve the probability of positive results. This is why systems thinking is so important because when you understand that business outcomes emerge from complex interactions between multiple system elements, you realize that trying to control specific results is often futile. However, you can influence the conditions that make desired outcomes more likely.

I observed this with a software company launching a crucial new product. The CEO coached the product director and his team to ensure that there was a carefully crafted stepwise approach that involved early adopter beta customers, brief intervals to test and learn, and fine-tuning of the functionality prior to the announcement. He also lobbied marketing, sales, and others to play well together. While some of his influence was based on positional authority (people have to listen), he didn't (couldn't) micromanage this. While the final product and go-to-market program was altered, his support and soft-gloved oversight ensured that the program succeeded. To reinforce what I talked about in the last chapter, the CEO applied the connection lens. He knew how various functions needed to work together, and the launch was too critical, so he focused on a few higher-impact areas where his influence could positively impact the outcome.

This shift, as I mentioned, may be a bit more complex to finesse. It requires accepting that in complex systems, direct control is often impossible, but intelligent influence is always available. Strategic leaders focus their energy on shaping conditions, such as culture, processes, and information flow, that enable success.

The influence mindset also changes how you might think about accountability. Instead of being accountable for specific results that may be outside your control, you assume accountability to create the best possible conditions for success. This approach allows for rapid learning so you can continually improve.

The paradox is that leaders who try to control everything often achieve less than those who focus on creating conditions for others to succeed.

TRY THIS:
For your next important initiative, identify your three biggest assumptions. Design small experiments to test those assumptions before committing to a full-scale plan.

FIVE PRACTICES TO BUILD YOUR STRATEGIC THINKING LATTICEWORK

Now that I've explained the fundamental shifts, I want to give you five practical approaches to build the mental models that enable strategic thinking under pressure. I call this building your latticework in honor of Charlie Munger's expression. Think of these practices as interconnected capabilities that reinforce each other, as shown in Figure 4.2. Each one makes the others more powerful. Like the interlocking elements of a strong framework, your strategic thinking capability becomes more robust as you develop all five practices together.

Figure 4.2 Five Practices to Build Your Strategic Thinking Latticework

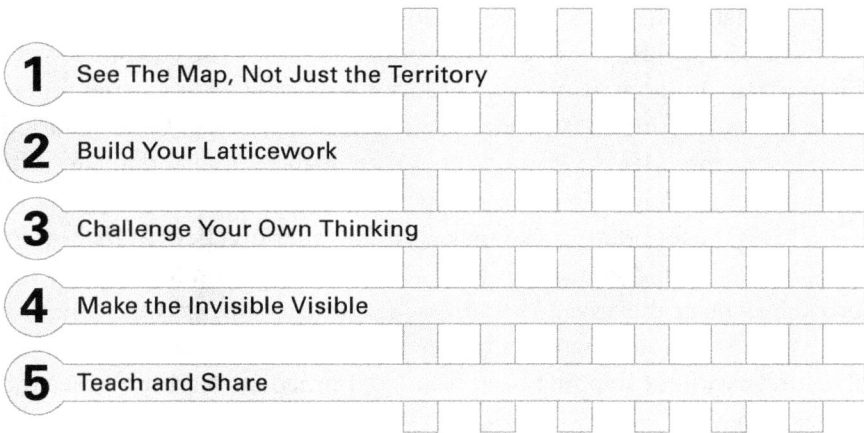

1. See The Map, Not Just the Territory
2. Build Your Latticework
3. Challenge Your Own Thinking
4. Make the Invisible Visible
5. Teach and Share

Practice 1: See the Map, Not Just the Territory

Most people get lost in the details of their immediate situation or the territory they can see right in front of them. Strategic thinkers learn to step back and see the map: the broader context, the patterns, and the forces that shape what's happening in their immediate view.

When you're dealing with a customer complaint, the territory is that specific issue. The map shows you whether this complaint represents a broader pattern of service problems, connects to recent product changes, or reflects shifting customer expectations in your market. When you're facing a budget shortfall, the territory is the numbers in front of you. The map reveals whether this reflects seasonal patterns, competitive pressure, or fundamental changes in your business model.

This practice directly enhances the pattern recognition habit from Chapter 2. When you regularly step back to see the bigger picture, you start noticing patterns that connect your immediate challenges to broader industry trends, market forces, and organizational dynamics.

This means regularly asking: What's the bigger picture here? How does this connect to other things happening in our industry, our company, our market? It means reading beyond your functional area, understanding how different parts of your business interconnect, and staying alert to forces that might not directly impact you today but could tomorrow.

I learned this lesson during my product management career when I realized that understanding the broader market map was just as important as understanding our specific product territory. The customers who bought our products were influenced by economic trends, competitive alternatives, and internal politics I had never considered. Once I started seeing these broader patterns on the market map, I could anticipate shifts months before they showed up in our sales data.

This is the foundation that makes all the other practices more powerful. You can't build effective mental models without understanding the broader context they need to address. You can't challenge your thinking effectively if you don't see the bigger picture. And you can't map invisible forces without stepping back from the immediate territory to see the larger systems at work.

TRY THIS:
Once a week, have a conversation with someone from a completely different department about how they approach their biggest challenges. Ask "What patterns do you see that others might miss?" Look for connections between their challenges and yours.

Practice 2: Build Your Latticework

Once you can see the map instead of just the territory, you realize that understanding complex business situations requires mental models from multiple disciplines. This goes back to Charlie Munger's concept that I mentioned earlier. To solve business problems, you need constructs that come from psychology, economics, systems theory, operations, and finance, because business problems rarely stay within one functional area.

When I was in college, I purposely took a second major in organizational psychology even though my primary major was management science. I had this feeling that I needed to understand people and culture because, from my internships, I recognized that people in larger organizations didn't always act rationally. I also studied economics because I wanted to understand market forces and operations research to learn about systems.

Here's what I want you to understand: your latticework becomes more powerful as you weave together insights from different domains. When you understand psychology, you can predict how people will respond to organizational changes. Or, you may better understand customer needs and motivations. When you understand economics, you can anticipate how market forces will affect your strategy. When you understand systems theory, you can see how different parts of your business influence each other.

But building your latticework isn't just about formal education. It's about actively seeking to understand how other functions, industries, and disciplines approach problems. When you have coffee with someone from finance, you're not just being social, you're adding economic and analytical thinking to your latticework. When you talk with customers about their business challenges, you're building market and operational insights.

The most effective approach I've found is to systematically expose yourself to four key domains that enhance strategic thinking: *human behavior* (psychology, organizational dynamics), *market forces* (economics, competitive dynamics), *systems and operations* (how things connect and flow), and *financial drivers* (what creates and destroys value). Each domain gives you a different lens for seeing the map more clearly.

This practice builds directly on seeing the map because the map becomes richer and more useful as you add different disciplinary perspectives to interpret what you're seeing. Your psychology knowledge helps you understand the human dynamics on the map. Your economics background helps you interpret market forces. Your systems thinking helps you see the connections and feedback loops.

TRY THIS:
Pick one area outside your current expertise that impacts your work. It could be finance if you're in operations, or customer behavior if you're in engineering. Spend 30 minutes this month having coffee with someone from that area. Ask them: "Help me understand how you think about [specific challenge]. What patterns do you see that I might miss from my perspective?" Look for ways their insights could apply to problems you're currently facing.

Practice 3: Challenge Your Own Thinking

The better you become at seeing the map and building mental models, the more dangerous overconfidence becomes. Success in strategic thinking can work against you. The more often you're right, the less likely you are to question your assumptions. Strategic thinkers deliberately seek out perspectives that make them uncomfortable.

This practice requires systematic approaches, not just good intentions. Here's what I've learned works:

First, actively seek out the smartest person you know who disagrees with your perspective on important issues. Schedule regular conversations with them. Don't try to convince them you're right. Use these discussions to stress-test your thinking.

Second, practice what I call "assumption archaeology." Before any important decision, write down your three biggest assumptions.

Then ask: What evidence supports this? What evidence contradicts it? What would I need to see to change my mind? This forces you to distinguish between what you know and what you're assuming.

Third, use the "pre-mortem" technique. Before implementing a strategy, imagine it's failed spectacularly. What went wrong? What assumptions were incorrect? What did you miss? This helps you identify blind spots before they become expensive mistakes.

Your cross-disciplinary latticework makes this practice more powerful because different mental models reveal different potential flaws in your thinking. Your psychology knowledge might reveal that you're underestimating resistance to change. Your economics perspective might show that you're ignoring competitive responses. Your systems thinking might reveal unintended consequences you hadn't considered.

The goal isn't to become paralyzed by doubt, but to hold your beliefs lightly enough that you can update them when new evidence emerges. Some of my best strategic insights have come from conversations with people who saw situations completely differently than I did, forcing me to examine assumptions I didn't even know I was making.

TRY THIS:
Next time you're confident about a decision, find someone who might see it differently such as someone from another department or with different experience. Ask them: "What am I missing here?" or "What would concern you about this approach?" Don't try to convince them you're right; just listen and see what you learn about your own assumptions.

Practice 4: Make the Invisible Visible

As you develop the ability to see the map and build your cross-disciplinary latticework, you start noticing that every organization has invisible forces shaping behavior and outcomes. These are the informal power structures, cultural norms, and systemic patterns that don't appear on any organization chart but profoundly influence how things get done.

Your expanded map view reveals these forces, but you need to actively identify and understand them. Early in my career, I was frustrated because many initiatives that seemed straightforward on paper ran into resistance or delays that didn't make logical sense. I started talking to managers in different departments to understand their perspectives and challenges. This taught me about resource constraints, competing priorities, and informal decision-making processes that were completely invisible from my original territory view.

I developed my organizational sensing to detect signals that revealed to me, how the organization really works versus how it's supposed to work. You might notice that certain types of initiatives always get delayed, even when they have formal support. Or you observe that people consistently interpret the same policy differently across departments. These patterns reveal invisible forces that can make or break your strategic initiatives.

Your latticework helps you interpret what these patterns mean. Psychology knowledge helps you understand why people resist certain changes. Economics thinking helps you see competing resource priorities. Systems awareness helps you identify where informal networks override formal processes.

Here's a practical approach: for any important initiative, map both the visible structure (org chart, formal processes, stated policies) and the invisible reality (who people go to for decisions, how information really flows, what unwritten rules govern behavior). The gaps between these two maps reveal the invisible forces you need to work with or change.

This practice is essential because strategic thinking without organizational reality is just academic exercise. You can have brilliant insights about the market map and powerful mental models, but if you don't understand the invisible forces in your own organization, your strategic initiatives will run into walls you never saw coming.

Making the invisible visible also strengthens your ability to challenge your thinking, because you start questioning not just your assumptions about markets and customers, but your assumptions about how your own organization really works.

TRY THIS:
For a week pay attention to informal conversations you have with others. For example, if they mention that they need to "run something by" someone, find out who that someone is, or whose opinion seems to matter most. Notice where the real decision-making occurs versus where it's supposed to take place according to the organization chart.

Practice 5: Teach and Share

This practice ties everything together and accelerates your development, which is why I eventually became a facilitator and trainer. It's because I discovered I love to teach. Even in my corporate life, I used to create "learning labs" where I'd bring people together from different functions to work through complex problems together, not just to solve them, but to understand how we were thinking about them.

When you teach others how you think through problems or challenges, you're forced to clarify your own mental models and identify gaps in your logic. I can't tell you how many times I've been explaining my approach to a problem and halfway through realized I was making assumptions I couldn't defend. Those learning labs taught me as much as they taught my colleagues.

Here's what I've discovered: teaching your thinking process isn't just good for others. It's one of the most effective ways to strengthen your own strategic thinking. When you explain your reasoning to people from different backgrounds, they ask questions that reveal blind spots you didn't know you had.

This creates a reinforcing loop that strengthens all your other practices. Teaching forces you to expand your field of view because you need to consider how others see the situation. It strengthens your mental models because you must explain connections between different disciplines. It challenges your thinking because others will question your assumptions. And it helps you map invisible forces because different people will point out patterns and dynamics you missed.

TRY THIS:
Next time you're working through a complex issue, reach out to a colleague and saying something like this: "I'm thinking through this problem and want to test my logic with you. Here's how I'm seeing it." State the issue, then ask: "What questions does this raise for you?" or "What am I not considering?"

These five practices work together systematically. Expanding your field of view reveals the need for multiple mental models. Building those models gives you frameworks for challenging your own thinking. Challenging your thinking helps you see invisible forces more clearly. And teaching others accelerates all of these capabilities while revealing new areas for development.

The goal isn't to master all five practices immediately, but to understand how they support each other. Start with whichever practice feels most relevant to your current challenges, but recognize that the real power emerges when they work together as an integrated system for building strategic thinking capability.

MANAGING YOUR MINDSET WHEN YOUR HAIR IS ON FIRE

It's easy to talk about these practices and how well they can work, but how do you put them into use when you're under pressure to perform and make decisions, often under tight deadlines? In a chapter on leadership in *The Product Manager's Desk Reference*, I suggested this: "Stay calm, even when your hair's on fire." When people are under pressure and the stakes seem high, it's easy to default to your own set mental models. Here's some advice that I've taken to heart and taught others to exercise:

- *Pause and breathe:* Before jumping to solutions, shift into learning mode. Ask What don't I understand about this situation?
- *Check assumptions:* Identify your immediate assumptions. Ask What if my first interpretation is wrong?
- *Consider context:* Think about what makes this situation unique. Ask "How is this different from similar situations I've handled before?
- *Focus on influence:* Identify what you can influence. Ask Where can I have the biggest positive impact?
- *Engage others:* Get different perspectives before deciding. Ask Who else should be thinking about this with me?

TRY THIS:
The next time you face a high-pressure situation, use this sequence: pause, check assumptions, consider context, focus on influence, engage others. Notice how this changes your response.

MAKING THE MINDSET SHIFT STICK

Developing a strategic mindset isn't a one-time change; it requires ongoing practice. Start with one practice and focus on it for a month. Create environmental cues to remind yourself to step back and see the bigger picture. Find colleagues who can help you practice these approaches. The key is creating systems that support strategic thinking rather than relying on willpower alone.

Leaders with a strategic mindset make better decisions, which creates better outcomes, which builds confidence in their approach. They become learning machines that get smarter with every challenge they face.

SUMMARY

A strategic mindset provides the mental foundation that makes all strategic thinking tools work when it matters most. The three fundamental shifts: from "knowing" to "learning," from "best practice" to "contextual thinking," and from "controlling outcomes" to "influencing conditions" transform how you approach complex challenges.

These shifts are supported by five interconnected practices that build what Charlie Munger called your strategic latticework: seeing the map instead of just the territory, building mental models from multiple disciplines, challenging your own thinking, making invisible forces visible, and teaching others to clarify your reasoning.

These five practices work together systematically. Expanding your field of view reveals the need for multiple mental models. Building those models gives you frameworks for challenging your own thinking. Challenging your thinking helps you see invisible forces more clearly. And teaching others accelerates all of these capabilities while revealing new areas for development.

The goal isn't to master all five practices immediately, but to understand how they support each other. Start with whichever practice feels most relevant to your current challenges, but recognize that the real power emerges when they work together as an integrated system for building strategic thinking capability.

By combining a strategic mindset with the five habits from Chapter 2 and systems thinking from Chapter 3, you develop the ability to see what others miss and create lasting value rather than temporary fixes.

You now have the habits that sharpen your thinking, the systems awareness that reveals hidden connections, and the mindset that embraces uncertainty as information. It's time to apply this foundation to one of the most challenging aspects of leadership: recognizing and solving problems before they become crises. In the next chapter, you'll discover "Step Zero" in problem-solving and learn to design solutions that address root causes while creating positive momentum throughout your organization.

5

STRATEGIC PROBLEM-SOLVING: FROM RECOGNITION TO SMART DECISIONS

Key Points

- Recognize problems worth solving before they become crises, and you'll spend energy on what matters most.
- Use variance analysis as detective work, and you'll spot patterns that reveal systemic issues that others miss.
- Design solutions that address root causes, and you'll prevent problems from recurring while strengthening your organization.

If I had an hour to solve a problem I'd spend 55 minutes defining the problem and 5 minutes solving it."

—ALBERT EINSTEIN

You now have the strategic thinking habits that sharpen your analytical capabilities, the systems awareness that reveals hidden connections, and the mindset that embraces uncertainty as information. These create your strategic thinking foundation. However, you'll see their true value when you can apply these in real situations.

In my opinion, problem recognition and problem-solving are vital aspects of strategic thinking. Why? Because most managers and leaders spend an inordinate amount of time reacting to problems that emerge suddenly. Their knee-jerk reactions seem out of proportion to the strategic issues that are linked to success. When I worked at Oracle, Larry Ellison, the CEO, would issue reactive directives that got filtered down through my boss, then to me, to then be channeled to my directors. Fire drills were the norm to the point that many of the employees reporting to the directors would complain that all they did was fight fires. I've been there, and so have you. What's worse, though, is that sometimes the problems we're supposed to fight have ripple effects elsewhere, so problems become magnified, not minimized.

What I've learned after decades of working with leaders who face this exact frustration is that, most of the time, the wrong problem is being solved. People rush to fix what they think is broken without asking why it broke in the first place. I see people who spend precious time on problems that don't matter, and they miss the problems that could derail their strategy.

Strategic problem-solving teaches you to spot the right problems early, frame them correctly, and design solutions that address root causes while strengthening your organization's capabilities. The difference isn't solving problems faster. It's solving different problems and addressing them in ways that prevent recurrence while creating positive momentum throughout your system.

UNDERSTANDING THE SIGNAL-TO-NOISE PROBLEM

Early in my career, I thought every variance, every complaint, every missed deadline required immediate attention. I was exhausting myself and my team chasing down issues that, in hindsight, didn't really matter. It took me years to understand what separates problems that deserve strategic attention from routine operational noise.

What Makes Problems Strategic?

The signal-to-noise concept comes from audio engineering, where engineers work to separate the signal (the music or voice you want to hear) from the noise, which is all the background interference, static, and distortion that gets in the way. A good audio system has a high signal-to-noise ratio, meaning the intended sound comes through clearly despite background interference.

Your business operates in the same way. In any business system, there's always "background noise." These are the routine operational variations that are normal and expected: a customer complaint here, a delivery delay there, a budget dispute somewhere else. These are just the natural variations that occur in complex systems.

However, buried within that noise are "signals." These are patterns that tell you something important is changing in your business system. When customer complaints start clustering around specific issues, when delivery delays correlate with certain suppliers, or when budget variances follow repetitive patterns across multiple departments, those are signals worth your attention.

The problem is that noise and signal can appear identical at first glance. Both show up as variances against expectations. Both might trigger the same alerts. Both might generate the same urgent emails demanding immediate action. The difference isn't in what happened, it's in what it means for your business system.

CHARACTERISTICS OF STRATEGIC PROBLEMS

Strategically significant problems have specific characteristics that separate them from routine operational issues:

- *They reveal system dynamics rather than isolated events.* When a few customers complain, those are likely unrelated events. When complaints spike in a pattern that correlates with product quality problems or delivery issues, they reveal a systemic issue that requires attention.
- *They connect to your competitive position or strategic capabilities.* A single customer complaint about poor service is operational. A pattern suggesting that service quality is affecting customer retention relative to competitors is strategic.
- *They recur despite short-term fixes.* If you keep solving the same type of problem repeatedly, you're probably treating symptoms rather than addressing the underlying system dynamic.
- *They affect multiple parts of your business simultaneously.* Strategic problems rarely stay contained in one department or function. They tend to ripple through an organization in ways that reveal interconnections.

TRY THIS:

This is designed to help you fortify your strategic problem recognition radar. For the next month, keep a simple "problem log" using the five habits from Chapter 2. When issues come to your attention:

- **Ask better questions:** Is this routine operational variance, a recurring tactical issue, or a potential strategic problem?
- **Look for patterns:** Does this connect to other issues you've noticed? What domain does it affect (market, operational, financial)?
- **Zoom out:** How does this fit into broader business trends you're tracking?
- **Learning mindset:** What would this teach you if it's signaling a bigger issue?
- **Time horizons:** Could this be an early indicator of something more significant?

Notice which category gets most of your time versus which should get most of your attention.

STEP ZERO: STRATEGIC PROBLEM RECOGNITION

Here's what I've observed about the best problem solvers: they don't necessarily solve problems faster than everyone else. Instead, they seem to spot the right problems earlier and understand which problems matter most. This capability qualifies for the term *step zero*, which is a general expression applied to a preparatory or initial action that must be taken before the start of any process. In this case, I'm referring to the problem-solving process. Step zero lets you spot strategic issues while they're still manageable, before they become the urgent crises that consume resources and limit your options. It's the difference between proactive solution design and reactive firefighting.

Using Strategic Habits for Problem Recognition

Remember the strategic thinking habits from Chapter 2? They transform how you recognize and prioritize problems. Better questions change everything in problem recognition. Instead of waiting for problems to announce themselves, strategic thinkers actively scan for them.

Instead of asking, What problems do we have?, ask, What patterns in our data don't match our assumptions about how our business works? Instead of asking, How do we fix this?, ask, What problem are we really solving, and why does it keep happening? Instead of asking, What's broken?, ask, What's working too well in one area that might be creating stress elsewhere?

Pattern recognition becomes crucial for strategic problem spotting. Strategic thinkers look for connections between seemingly unrelated issues. When customer satisfaction scores remain stable, while customer acquisition costs gradually increase, that pattern suggests an issue of market dynamics. When employee engagement drops, while productivity temporarily improves, that pattern might predict future performance problems.

The Communication Problem Trap

I see this trap all the time, especially when teams convince themselves they have a "communication problem." When I ask detailed questions about these situations, I hear responses that suggest people aren't sharing information effectively or deadlines are being missed. It's as if people who work together point fingers at one another when things go wrong.

I've learned to ask a different question: What did you want to have happen? Sometimes employees can talk about the goal that wasn't met. Then I can ask, Was everyone clear on what they were supposed to do, by when, and with whom? From these conversations, I've learned that employees needed to clarify goals, roles, responsibilities, and timing. They use the blanket term *communication problem* when they're seeing symptoms without understanding the underlying system that creates those symptoms.

TRY THIS:

Use this exercise to apply systems thinking to communication problems. The next time your team identifies a "communication problem," use the systems lenses from Chapter 3:

- **Connection lens:** Map what this issue connects to: goals, roles, processes, resources
- **Assumption lens:** What did we assume would happen that didn't?
- **Time lens:** When did the breakdown occur, and what delays might be involved?

Ask these strategic questions: What do we really think is going on? What if this isn't really a communication problem? Then go to: What did we want to have happen that didn't? This helps you dig through symptoms to find root causes.

VARIANCE ANALYSIS: YOUR STRATEGIC DETECTIVE TOOL

This is where the detective work begins, and it's one of the most powerful tools I know for developing the strategic thinking needed to optimize your effectiveness. Variance analysis isn't just looking at performance versus targets; it's recognizing that variances tell stories about what's happening in your business system.

Patterns Tell Stories

Similar to my definition of a problem, a variance is a gap between a "plan" and an "actual." You see these in financial reports or when comparing key performance indicators to targets. When I'm analyzing financial statements, I look at the plan, the actual, the "delta" or difference, and the delta as a percentage of the plan. But here's what I've learned: even when you're looking at a single variance, you won't understand what's really happening. Why? Because strategic thinkers know that variances don't happen in isolation.

Instead, they connect to one another, and there's generally a rippling impact across the areas being studied. This compounding tends to reveal patterns that point to deeper systemic issues and, ultimately, potential solutions.

Think about it this way: if sales drop 5 percent, that might not mean much by itself. But if sales drop 5 percent, while customer satisfaction remains stable, customer acquisition costs increase 12 percent, and sales cycle length extends by 20 percent, that combination tells a story about fundamental changes in your competitive landscape that require a strategic response.

Connected Variance Analysis

Most successful leaders have an uncanny ability to process multiple data streams simultaneously and look for connections between seemingly unrelated variances. They don't just ask, Why did sales drop 5%? They ask, What story do these connected variances tell us about our business system?

Here's a pattern I see repeatedly: companies notice declining profitability and immediately focus on cost reduction. But when you connect the variances, you might discover that profitability dropped because customer acquisition costs increased, which happened because customer lifetime value decreased, which occurred because retention rates fell due to service quality issues. The real solution isn't cost cutting; it's fixing the service quality problem that's driving the entire cascade.

TRY THIS:
For your next business review, apply strategic thinking systematically:

- **Pattern recognition:** Pick three problematic metrics and map how they connect
- **Better questions:** Instead of asking "Why did X happen?" ask "What story do these connected variances tell us about our business system?"
- **Systems thinking:** Use the feedback lens—are these variances creating reinforcing loops or revealing system resistance?
- **Time horizons:** What might these patterns predict about future performance?

Document your findings because this becomes a vital part of your strategic intelligence.

STRATEGIC PROBLEM PRIORITIZATION

One of the biggest challenges in problem recognition is distinguishing between urgent problems and important problems. I learned this the hard way early in my career when I was constantly firefighting urgent issues, while the truly important work got postponed.

Breaking the Urgency Addiction

Some of us are wired to respond with equal urgency to various situations. You might relate to this when you rush to answer a call or react to an angry email. It seems as if every situation could be seen as a crisis, and soon we're numb to the differences and we become conditioned to knee-jerk reactions. I've seen this in companies when a boss gets a call from an angry executive about a problem the executive thinks is catastrophic, and everyone drops everything to fix it. No one asks questions because that would seem like insubordination. No one just says, "Wait a second; let's figure this out." The result is that we get conditioned to this type of behavior. This is not productive for strategic thinkers.

The Strategic Sweet Spot

Many problems that deserve attention fall into what Stephen Covey, the author of *The 7 Habits of Highly Effective People*, called "important but not urgent." We might see a shift in customer buying behavior. We might observe a person who doesn't fulfill commitments to project teams. We might learn that our product launches are consistently late. These might be ignored as organizational noise. It's that numbness I just mentioned.

However, these situations offer you a chance to examine things more closely, to step back and process what's happening that leads to these situations. Those who do may spot areas that can be addressed without knee-jerk reactions and offer insights that can keep an important item from becoming urgent, while addressing the root cause. The main idea is to deal with issues with strategic depth, not tactical speed.

THE URGENCY IMPORTANCE MATRIX

Most of us are challenged when it comes to prioritizing—deciding the order of importance or urgency for problems and issues. In my workshops this issue comes up more than 90 percent of the time. There are many ways to address this, but one that has broad implications for step zero, as well as overall problem recognition, is the urgency importance matrix. If you haven't heard of it, let me share the backstory with you. President Dwight Eisenhower had a time management philosophy. It was said that he was often faced with a barrage of seemingly urgent demands, and he needed a way to prioritize things. He once said: "What is important is seldom urgent, and what is urgent is seldom important." This matrix was also popularized by Stephen Covey, who labeled it a time management matrix.

This can be a helpful tool for prioritizing problems, issues, and challenges. It can help you balance doing things you deem important and doing things that are good for the system of the business. The urgency importance matrix is shown in Figure 5.1.

Figure 5.1 The Urgency Importance Matrix

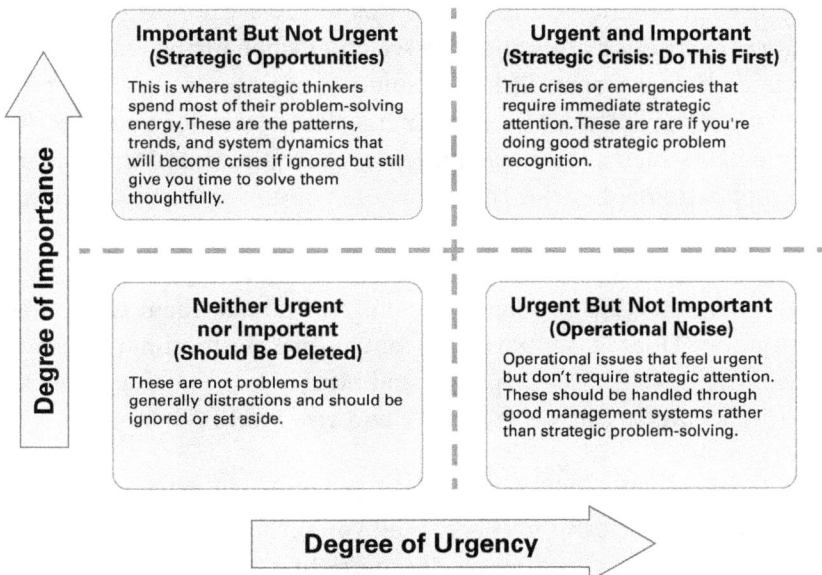

Important But Not Urgent (Strategic Opportunities)
This is where strategic thinkers spend most of their problem-solving energy. These are the patterns, trends, and system dynamics that will become crises if ignored but still give you time to solve them thoughtfully.

Urgent and Important (Strategic Crisis: Do This First)
True crises or emergencies that require immediate strategic attention. These are rare if you're doing good strategic problem recognition.

Neither Urgent nor Important (Should Be Deleted)
These are not problems but generally distractions and should be ignored or set aside.

Urgent But Not Important (Operational Noise)
Operational issues that feel urgent but don't require strategic attention. These should be handled through good management systems rather than strategic problem-solving.

Degree of Importance

Degree of Urgency

TRY THIS:
Categorize your current "problems" using this framework. Ask yourself: What percentage of your problem-solving time is spent in each quadrant? Where should it be spent for maximum strategic impact? If you wish, you might apply a scoring mechanism. For example, in degrees of importance, you can rank them on a 1–10 scale and similarly for degree of urgency. This might offer you some perspectives that will allow you to prioritize items, and perhaps, make appropriate trade-offs.

The key insight is that most strategic problems start in the "Important But Not Urgent" category. By the time they become urgent, your strategic options are limited and often more expensive.

DECISION MATRIX: MAKING TRADE-OFFS VISIBLE

A decision matrix is a simple tool that helps you compare multiple options against the criteria that matter most for your situation. Instead of following your instinct or picking what seems best, you systematically evaluate each option. You could use this tool, for example, for product feature prioritization and candidate selection when hiring employees.

Here's how it works: List your solution options across the top of a grid. Down the left side, list the criteria that matter for your decision. Include items such as strategic fit, resource requirements, timeline, risk level, and customer benefit. Then score each option against each criterion using a simple scale of 1 (poor fit) to 10 (excellent fit).

For example, if you're choosing between three approaches to solve a customer retention problem, you might evaluate them on strategic importance (Does it support our competitive positioning?), resource availability (Do we have the people and budget?), speed of implementation (How quickly can we execute?), and risk level (What happens if it doesn't work?).

Add up the scores for each option. The highest total score indicates which solution performs best across all your criteria. But here's the key: the highest score isn't automatically the right choice. The matrix helps you see the trade-offs clearly. Maybe option A scores highest overall but performs poorly on the timeline criterion, whereas option B scores slightly lower but can be implemented much faster. The real value is making your thinking visible and forcing yourself to consider multiple factors instead of just one or two that seem most obvious.

Figure 5.2 is a template for a decision matrix that you can use both for prioritization and problem-solving.

Figure 5.2 Decision Matrix Template

	Strategic Importance	Deliver Better Value to Customer	Provide Competitive Advantage	Total Score
Solution Idea 1				
Solution Idea 2				
Solution Idea 3				
Solution Idea 4				

AI AS YOUR STRATEGIC PROBLEM DETECTION SYSTEM

The most powerful application of AI in problem-solving isn't in the generation of solutions, it's finding problems worth solving before they become crises. AI excels at the pattern recognition and variance analysis that makes "Step Zero" systematic rather than intuitive.

Automating Your Strategic Radar

Remember the signal-to-noise challenge from earlier in this chapter? AI systems can be taught to continuously monitor select business metrics and flag variance combinations that historically preceded strategic problems. Instead of waiting for quarterly business reviews to reveal troublesome trends, AI can alert you when subtle pattern changes suggest emerging issues.

For example, train an AI system to monitor the variance patterns we discussed: stable customer satisfaction plus rising acquisition costs plus lengthening sales cycles. When this combination appears, AI flags it as a potential problem (perhaps months) before these impact reductions in revenue.

Enhanced Variance Analysis Detective Work

AI transforms variance analysis from periodic detective work into continuous intelligence gathering. Teach your AI system enough historical variance data, along with the outcomes that followed so it can learn to recognize which variance combinations predict strategic challenges versus operational noise.

The key is teaching AI your strategic thinking framework for problem recognition. Here are some ideas for you to consider:

- **Pattern Recognition Training:** Show AI examples of variance patterns that revealed strategic problems in your business
- **Systems Connection Mapping:** Train AI to flag when problems in one area correlate with changes in seemingly unrelated areas
- **Time Horizon Analysis:** Teach AI the typical lag times between early warning signals and strategic impact in your industry

Building Your AI Problem Recognition Partnership

Step 1: Document Your Problem Recognition Process. I strongly suggest that you and your team take time to Write down how you currently spot strategic problems:

- What variance combinations do you investigate?
- Which metrics do you watch most closely?
- What early warning signals have predicted problems in the past?

Step 2: Train AI on Historical Patterns. Feed your AI system examples of:

- Variance patterns that preceded strategic challenges
- False alarms that looked concerning but weren't strategic issues
- The timeframes between early signals and actual problems

Step 3: Create Smart Alerts. Instead of generic threshold alerts, design AI prompts like:

- "Flag variance combinations similar to the Q2 2023 customer retention crisis"
- "Alert me when you detect patterns that historically preceded competitive threats"
- "Monitor for early indicators of the supply chain issues we faced last year"

The Strategic Advantage

This focused method gives you three to six months' advance warning on strategic challenges. While competitors react to problems after they become obvious, you're able to address issues while they're still manageable. AI doesn't replace your strategic judgment about what problems mean. Instead, it ensures you never miss the early signals that strategic problems are developing.

The goal isn't perfect prediction, but systematic early detection that gives you time for strategic response rather than reactive firefighting.

THE STRATEGIC PROBLEM-SOLVING PROCESS

When you've determined that there's a problem worth addressing, I believe a systematic process can help organize your thinking. This process builds on the patterns you've surfaced and your problem-solving instincts. When this is underscored with strategic thinking rigor, you'll end up with solutions that you and your team can support.

Step 1: Reframe the Problem Using the Strategic Lens

You've already recognized this as a problem worth solving. Now systems thinking becomes crucial. The problem you initially spotted is often a symptom of a deeper systemic dynamic. Reframing is like using active listening in communication. When someone tells you something, you might reply with, "So what you're saying is . . . " You're using enhanced questioning to reframe and clarify the real issue.

Ask questions such as

- What happened that wasn't supposed to happen?
- What didn't happen that should have happened?
- Did it challenge our assumptions about how things work?
- Did it impact customers in ways that affect our competitive position?
- Did it reveal gaps in our systems or capabilities?
- What does this suggest about our business model or strategy?

TRY THIS:

Write three different problem statements for the same issue: one focused on the immediate symptom, one focused on the underlying pattern, and one focused on the system dynamic that creates the pattern. Notice how different framings suggest different solutions.

Step 2: Validate Your Understanding

Before jumping to solutions, make sure you really understand what you're dealing with. Think of this like a doctor confirming a diagnosis before prescribing a treatment.

Ask yourself: Can I clearly explain this problem and its causes to someone else? What would they ask that I can't answer confidently?

If you can't explain it clearly, you may need to

- Talk to people closer to the problem
- Get customer or stakeholder perspectives

- Understand how this connects to other business issues
- Clarify the financial or operational impact

TRY THIS:
Explain the problem and its causes to someone not involved with the situation. If they ask questions that you can't answer confidently, those are your knowledge gaps. Address the critical ones before moving to solutions.

Don't spend weeks gathering more data. Just fill in the knowledge gaps that would change how you approach the solution.

Step 3: Generate Strategic Options

Don't jump to the first solution that comes to mind. Generate multiple options, then see if you can combine the best parts of different approaches.

Start by asking What are all the different ways we could approach this? Think about solutions that

- Fix the immediate problem quickly
- Address the root cause permanently
- Prevent similar problems in the future
- Turn this challenge into an opportunity

Write down every option, even ones that seem impractical initially. Then ask these questions:

- Is there any reason we can't do multiple approaches?
- Can we find a way to resource them all?
- Do they interfere with each other or create synergies?
- What's the additional cost of doing the second option alongside the first?
- What's the additional benefit of combining approaches?

Sometimes the best solution combines immediate fixes with long-term improvements, or tactical responses with capability building.

TRY THIS:

Pick a current problem that's been nagging your team. First, spend 10 minutes reframing it using the three-statement approach: write one problem statement focused on the symptom, one on the underlying pattern, and one on the system dynamic creating the pattern. Then, generate at least five solution options without judging them. Set a timer for 15 minutes and go. Next, create a simple decision matrix with 4–5 criteria that matter for your situation (strategic impact, resource requirements, timeline, risk level, etc.). Score each solution 1–10 against each criterion. Finally, look for combination opportunities: Can you address the immediate symptom while building long-term capabilities? Can you pilot one approach while preparing another? Often the best strategic solution combines quick wins with systemic improvements. Document your process—what patterns did you notice about how you naturally frame problems? Which criteria mattered most in your decision? This becomes part of your strategic problem-solving toolkit.

Step 4: Evaluate and Choose Your Solution

Now compare your options using criteria that matter for your business. Don't just pick what feels right initially. Think through the trade-offs systematically first. This is where a decision matrix helps you compare options against criteria that matter.

Decision criteria may include

- Does it impact our strategic goals and positioning?
- Will it improve customer experience or competitive advantage?
- Do we have the resources and capabilities to execute it well?
- Does it help us become more efficient or effective?
- What's the risk if it doesn't work as planned?

> **TRY THIS:**
> Create a simple decision matrix with each solution scored against the criteria that matter most (1–10 scale works fine). Don't just use generic criteria—derive decision factors that make sense for your specific situation. The highest score isn't always the right choice, but it helps you understand the trade-offs.

DECISION-MAKING UNDER UNCERTAINTY

Here's the reality: you'll rarely have all the data you'd like when facing business problems. You could spend months gathering more information, but by then the problem might have evolved, or the opportunity might be gone. Strategic thinkers learn to make good decisions with incomplete information.

The Paleontologist's Approach to Strategic Decision-Making

I live in New York City, and I've been going to the American Museum of Natural History for years. What fascinates me about paleontologists is how they approach incomplete information, which is a challenge every strategic thinker faces daily.

When paleontologists find a T. rex skeleton, they never find all the bones. They might have the skull, most of the spine, and some leg bones, but they're always missing pieces such as ribs, part of the tail, vertebrae. Here's what they don't do: they don't wait decades hoping to find the exact missing pieces. Instead, they use clay models to fill the gaps.

They study the bones they have, examine similar dinosaurs, understand how bone structures work, and create hypotheses about what the missing pieces should look like. Their clay models represent informed assumptions based on the best available evidence. Crucially, they label which parts are actual fossils and which are clay reconstructions.

Your Strategic Paleontology Toolkit

This is exactly what you need to do with strategic problems. You use the data you have (your fossil bones), apply what you know about similar situations (comparative analysis), understand how business systems work (your structural knowledge), and make reasonable hypotheses about the missing pieces (strategic assumptions).

The key is being explicit about what you're assuming versus what you know for certain, just as paleontologists distinguish fossils from clay. Here's how to apply this systematically:

Identify Your "Fossil Bones" (What you know for certain):
- Verified financial data and performance metrics
- Confirmed customer feedback and market research
- Documented competitive actions and market changes
- Proven causal relationships from prior analyses

Map Your "Missing Bones" (Critical information gaps):
- Customer motivations behind behavior changes
- Competitive response timing and intensity
- Economic factors that could affect demand
- Internal capability requirements for success

Create Your "Clay Models" (Informed hypotheses): Using your strategic thinking framework:
- *Pattern recognition:* What do similar situations suggest about missing pieces?
- *Systems thinking:* How would different assumptions affect system dynamics?
- *Cross-discipline thinking:* What do psychology, economics, or operations suggest?
- *Time horizon analysis:* What assumptions make sense given realistic timelines?

Label Your Assumptions Clearly: Document which parts of your analysis are based on solid data (fossils) versus reasonable assumptions (clay). This prevents you from treating hypotheses as facts and helps you stay alert to new information that could change your clay models.

The Strategic Advantage

Just as museums display magnificent dinosaur skeletons that combine fossils with clay reconstructions, you can make excellent strategic decisions by combining solid data with well-reasoned assumptions. The paleontologist's discipline about being explicit about what's known versus assumed prevents you from either waiting forever for complete information or making decisions based on wishful thinking.

This approach transforms uncertainty from a paralysis-inducing problem into manageable strategic intelligence. You're not guessing. You're making informed hypotheses that you can test and refine as new evidence emerges.

Trusting Strategic Intuition

Sometimes the analysis points one way, but your experience suggests something different. Don't automatically dismiss that feeling, but don't let it override systematic analysis either. Strategic intuition works best as the final layer of your decision-making process, after you've done the analytical work, gathered data, used decision matrices, and understood the trade-offs systematically. Don't automatically dismiss that feeling. Your strategic intuition isn't magic or guesswork. It's your brain processing patterns from years of experience faster than your conscious mind can analyze them. It's the accumulated wisdom from seeing similar situations play out over time.

When you've seen customer behavior shifts before, part of you recognizes the pattern even when the current data look different. When you've worked through similar competitive situations, you develop instincts about what works and what doesn't. When you've launched products or managed teams through challenges, you build a library of analogous experiences that inform your judgment. These are often below the level of conscious awareness.

I've seen leaders make poor decisions because they ignored their instincts in favor of what looked good on paper. I've also seen leaders make excellent decisions that couldn't be fully justified analytically but drew on deep experience with similar situations. The key is understanding when to rely on this experiential wisdom.

Here's how to use strategic intuition effectively: Complete your systematic analysis first. Understand the data, evaluate the options, and think through the trade-offs. Then ask yourself: Given everything I know about situations like this, does this decision feel right? If something feels off, dig deeper. What pattern from your experience is triggering that concern? What question haven't you asked yet?

Strategic intuition becomes your quality check after rigorous analysis. It helps you choose between analytically equivalent options or alerts you when you might be missing something important. It's not about making snap judgments, it's about integrating analytical thinking with experiential wisdom to make better strategic decisions.

BUILDING STRATEGIC PROBLEM-SOLVING CAPABILITIES

The problem-solving capabilities you develop through strategic thinking become essential for the strategy formulation work we'll explore in Chapter 6. Strategy formulation is fundamentally about solving complex, interconnected problems: How do we compete successfully in changing markets? How do we build capabilities faster than competitors? How do we allocate limited resources for maximum strategic impact?

Integration with Your Strategic Foundation

The variance analysis skills help you understand your dynamic baseline and spot signals that your strategic position is shifting. The pattern recognition capabilities help you synthesize complex insights and identify strategic themes that others miss. The systems thinking helps you design solutions that create reinforcing advantages rather than isolated improvements.

The strategic questioning approaches help you work backward from desired outcomes to identify what problems you need to solve to achieve strategic success. The solution development frameworks help you design initiatives that build capabilities while delivering results.

Most importantly, the strategic mindset helps you approach problem-solving as continuous learning rather than one-time fixes. You recognize that strategic success requires constantly sensing, interpreting, and responding to complex, evolving challenges in ways that create sustainable competitive advantage.

DEVELOPING ORGANIZATIONAL INTELLIGENCE

People who consistently apply strategic thinking to problem-solving don't just solve problems better; they build problem-solving intelligence that gets stronger over time. Each problem solved strategically builds capabilities that serve the organization across future challenges.

People in these firms deploy resources or utilize systems to spot problems earlier because they've trained multiple people or systems to recognize strategic signals. They solve problems more effectively because they've developed systematic approaches to complex challenges. They prevent problem recurrence because they address root causes rather than symptoms.

This is what separates truly strategic organizations from those that just have good strategic plans. They've built the capability to continuously identify and solve the problems that matter most for competitive success.

SUMMARY

1. *Strategic problem recognition separates proactive leaders from reactive "firefighters."* Understanding signal versus noise lets you spot strategic issues while they're manageable, not after they become crises that limit your options.
2. *Variance analysis reveals the hidden stories your data are telling.* Connected patterns across multiple metrics expose fundamental shifts in competitive dynamics, customer behavior, or market conditions that require strategic responses, not tactical fixes.
3. *Focus energy on "important but not urgent" problems.* Most strategic issues start in this quadrant. By the time they become urgent, your options are limited and expensive. Strategic thinkers prevent crises rather than just managing them.
4. *Reframe problems before generating solutions.* The problem you initially spot is usually a symptom. Strategic reframing using systems thinking ensures that you're solving the right problem, not just the visible one.

5. *Generate multiple options, then combine the best approaches.*
   Strategic problem solvers don't settle for the first solution. They
   explore various approaches and often find that combining
   immediate fixes with long-term improvements creates the
   strongest solutions.

6. *Make good decisions with incomplete information.* Like
   paleontologists filling gaps in a dinosaur skeleton with clay,
   strategic thinkers use hypotheses based on the best available
   evidence. The key is being explicit about assumptions and
   staying alert to new information.

7. *Build AI partnerships that amplify your strategic thinking.* AI
   excels at pattern recognition and data processing, but strategic
   judgment about what patterns mean and what actions to take
   remains uniquely human. Train AI to enhance your capabilities,
   not replace them.

You now have the complete strategic thinking foundation: habits that sharpen your analysis, systems awareness that reveals hidden connections, mindset that embraces uncertainty as information, and problem-solving capabilities that address root causes. These aren't just individual skills. They're the integrated platform for the ultimate strategic thinking application.

In Chapter 6, I want to help you to pull it all together to transform how you create, or influence strategies that actually work in dynamic business environments. You'll discover how strategic thinking converts strategy formulation from template completion to strategic insight generation, from annual planning events to continuous competitive advantage building.

Everything I've shared, including pattern recognition, systems lenses, strategic questioning, variance analysis, etc, offer you the tools in your toolbox to formulate strategies that adapt and thrive rather than break when conditions change. The most effective strategies don't come from following frameworks; they emerge from applying strategic thinking systematically to the fundamental questions every business must answer.

It's time to become a strategic architect of your organization's future.

6

STRATEGIC THINKING MEETS STRATEGY FORMULATION: TRANSFORMING PROCESS INTO POWER

Key Points

- Turn strategy formulation from template completion into strategic insight generation that creates real competitive advantage
- Apply everything you've learned to create strategies that adapt and thrive rather than break when conditions change
- Integrate all your strategic thinking capabilities to build strategies that work in the real world, not just on paper.

The art of strategy is not in following the process perfectly, but in thinking strategically while you're in the process.

—STEVEN HAINES

You've built the strategic thinking foundation through five core habits, developed systems awareness that reveals hidden connections, cultivated the mindset that embraces uncertainty as information, and mastered problem-solving approaches that address root causes. Now comes the ultimate application: transforming how you create strategies that actually work in dynamic business environments.

This chapter isn't about giving you another strategic planning framework. Instead, it's about showing you how strategic thinking transforms the fundamental questions every business must answer:

1. Where have we been?
2. Where are we now?
3. Where should we go?
4. How will we get there?
5. How will we know we're succeeding?

The questions don't change. How you think about them changes everything.

THE STRATEGIC THINKING TRANSFORMATION: BEYOND PROCESS TO STRATEGIC INTELLIGENCE

I've been there. We've all been there. Your team spends weeks in strategic planning sessions, follows every best practice, and creates comprehensive slide decks with market analysis and competitive assessments. Your customer research seems flawless. Your competitive analysis gives you the proof that you can win. Your options and initiatives, while ambitious, seem attainable. What a great process!

Six months later, you're wondering why nothing has really changed. The strategy deck sits in a file on some shared server while everyone goes back to fighting the same fires, solving the same problems, and getting the same results they got before the planning sessions.

Here's what I've learned after decades of watching smart leaders follow strategic planning processes by the book and still end up with strategies that don't work: the difference isn't in the process. It's certainly in the quality of data, but even more, in the thinking that happens during the process.

THE STRATEGIC COACH ANALOGY

Think about the difference between a novice football coach and an experienced strategic coach facing the same situation in a game. Both know the basic process: assess field position, consider down and distance, call a play. But the strategic coach brings something fundamentally different to that process.

The strategic coach uses pattern recognition (Chapter 2) to see how current conditions connect to situations they've observed before. They apply systems thinking (Chapter 3) to understand how their play call will affect not just this down but create ripple effects throughout the entire game dynamic. They demonstrate the learning mindset (Chapter 4), constantly updating their understanding based on what the other team does. They think across time horizons (Chapter 2), balancing immediate needs with longer-term game management.

The same transformation happens when you bring strategic thinking to strategy formulation. You're still answering the fundamental questions, but you're thinking about them in ways that create dynamic, adaptive, insight-driven strategies instead of static plans.

This difference shows up immediately in how strategic thinkers approach fundamental questions every strategy must answer: Where have we been, and Where are we now? Most leaders treat this as a historical exercise where they document past performance and current position. Strategic thinkers treat it as detective work that reveals the patterns and momentum shaping their business reality.

YOUR DYNAMIC BASELINE: THE STATE OF THE SYSTEM

Every strategy formulation process starts with understanding Where have we been? and Where are we now? But here's where strategic thinking creates the first major transformation: you stop treating your baseline as a static snapshot and start seeing it as a constantly moving line of scrimmage. (A *line of scrimmage* in American football is simply the imaginary line across the field where the ball currently sits at the start of each play. It's your team's exact position at any given moment and your starting point for the next move.)

Think of your business baseline as the line of scrimmage in football. At any given moment, you know exactly where you are: your financial position, market share, customer segments, operational capabilities, competitive standing, and organizational health. This is your current line of scrimmage. It's your starting point to consider your next strategic move.

But here's the crucial insight that separates strategic thinkers from traditional planners: this line shifts constantly. Every quarter's financial results move the line. Every competitive announcement moves the line. Every customer behavior shift, every operational improvement, every market change moves your strategic line of scrimmage to a new position.

Understanding that your baseline constantly shifts creates a new challenge: how do you track and analyze this movement at the scale and speed modern business requires? This is where artificial intelligence (AI) becomes essential.

AI AS YOUR STRATEGY FORMULATION ACCELERATOR

The future of strategy formulation isn't choosing between human insight and artificial intelligence, it's combining your strategic thinking capabilities with AI's pattern recognition and processing power to create strategies that adapt faster and perform better than either could achieve alone.

While AI shouldn't replace the judgment you need to call upon to interpret what patterns mean for your unique context, it can dramatically enhance every element of strategy formulation I've discussed: dynamic baseline tracking, SWOT synthesis, strategic option evaluation, and resource allocation optimization.

The Strategic AI Partnership Model

I like to think of AI as a research analyst with superhuman pattern recognition capabilities but no business context. It can process thousands of strategic decisions, market shifts, and competitive moves to identify patterns you'd never spot manually. *It needs your strategic thinking logic to understand what those patterns mean and how to apply them to your specific situation.*

This partnership works best when you combine AI's strengths, which include data processing, pattern recognition, and scenario analysis with your strategic thinking capabilities (contextual interpretation, systems awareness, and strategic judgment). Neither replaces the other; together they create competitive advantages neither could achieve alone. I cannot emphasize this enough. Do not lean on the AI machine to make judgments and decisions about your strategic intent.

I'd like to offer you some suggestions where you can aim your AI training. Specifically, areas where pattern recognition at scale has a chance to surface strategic advantages. These include: understanding your shifting competitive position, evaluating strategic options against thousands of comparable decisions, and optimizing resource allocation based on strategic outcomes rather than historical patterns. I have some ideas for you to apply this partnership model to the three areas where, in my belief, AI can have an outsized impact in your strategy formulation process:

Three Strategic AI Applications

Application 1: Dynamic Baseline Intelligence. Your "line of scrimmage" constantly shifts based on competitive moves, market changes, and internal developments. AI can monitor these shifts continuously rather than waiting for quarterly reviews.

Train AI to track your dynamic baseline by feeding it:

- Historical performance data with contextual narratives about what caused major shifts
- Competitive analysis information, including competitive products and actions of competitive companies
- Assessments of market environment shifts (e.g., politics, regulations, technology, etc.)
- Information about your company's staffing and human resource expertise

Use questions like these to prompt your AI system to see if it reveals the intelligence you need.

- "What variance patterns in our metrics suggest our competitive position is shifting?"
- "Based on similar companies' data, what early indicators suggest market disruption ahead?"
- "How do current baseline shifts compare to patterns that preceded our major strategic pivots?"

The intended result: Instead of discovering strategic position changes months after they've occurred, you get early warning signals that let you adapt proactively. In my experience, I recommend a trial-and-error, or test-and-learn approach to fine-tune how the system responds to your queries and to determine where you might need to augment your data inputs.

Application 2: Strategic Option Analysis at Scale. Traditional strategic option evaluation relies on your team's experience with similar decisions. You can leverage AI to analyze thousands of comparable strategic choices across industries and contexts to enhance your multi-lens evaluation. Here are some examples:

- Identify similar strategic moves by comparable companies and the associated outcomes
- Predict resource requirements and timeline challenges based on historical patterns
- Flag potential unintended consequences by analyzing comparable decisions
- Suggest option combinations that historically created synergistic effects (which is really part of the continual training of your AI system)

Strategic option prompts you can try:

- "What were the outcomes when companies similar to ours pursued this type of strategic option?"
- "What capability gaps typically emerge when implementing strategies such as ours, and how long do they take to address?"
- "What combination of strategic options has historically created the strongest competitive advantages in our industry?"

Integration with your multi-lens approach: AI provides the pattern recognition across thousands of examples, while you apply the five lenses to interpret what those patterns mean for your specific context, culture, and competitive position.

Application 3: Resource Allocation Optimization. Remember the variance analysis approach to resource allocation where you can examine gaps between stated priorities and actual resource deployment? As long as it has the right data input or linkages to some of your data systems, AI can continuously monitor these patterns and optimize allocation based on strategic outcomes rather than historical budgeting.

AI can often be taught to analyze:

- Resource allocation patterns vs. strategic outcome attainment
- Cross-functional resource dependencies that constrain strategic options
- Portfolio effects where resource allocation in one area impacts others
- Optimal sequencing for capability investments based on strategic option priorities

Here are some resource optimization prompts you might try:

- "What resource allocation patterns historically produced the best strategic outcomes for companies like ours?"
- "Where are we systematically under-investing relative to our stated strategic priorities?"
- "What resource reallocation would create the biggest strategic impact given our current baseline and strategic options?"

Building Your Strategic AI Capability

Phase 1: Document Your Strategic Intelligence (Month 1). Create AI training materials by documenting:

- Your enhanced SWOT process and the insights it typically generates
- Historical examples of successful and failed strategic decisions with context
- Your working backward planning processes and milestone tracking
- Resource allocation decisions and their strategic outcomes

Phase 2: Train AI on Your Strategic Context (Months 2-3). Begin feeding AI your documented processes along with:

- Industry-specific strategic challenges and successful responses
- Your company's unique competitive context and stakeholder considerations
- Historical variance patterns that preceded strategic successes or failures

Phase 3: Integrate AI into Strategy Formulation (Month 4+). Start using AI to enhance each element of your strategy formulation process:

- Baseline analysis: AI pattern recognition + your strategic interpretation
- Synthesis using SWOT: AI cross-industry examples + your contextual analysis
- Option or opportunity evaluation: AI outcome analysis + your multi-lens assessment
- Resource optimization: AI allocation modeling + your strategic judgment

The Strategic Advantage Principle

Organizations that accelerate their human-AI strategic partnerships have a chance to outperform those that view AI as either a replacement for strategic thinking or a peripheral tool. The competitive advantage comes from leaders who understand strategic thinking well enough to direct AI effectively and interpret AI insights through strategic frameworks.

Your investment in strategic thinking development, including the five habits, systems awareness, strategic mindset, and problem-solving capabilities, becomes more valuable, not less, in an AI-enhanced world. Someone still needs to provide the context, judgment, data, and interpretation that turns AI processing power into strategic advantage.

The leaders who will dominate the next decade aren't those who predict which AI tools will emerge, but those who develop the strategic thinking capabilities to leverage any AI advancement effectively.

SWOT AS THE ENGINE OF STRATEGIC SYNTHESIS: BEYOND BRAINSTORMING TO STRATEGIC INTELLIGENCE

Now we get to where strategic thinking creates the biggest transformation in strategy formulation. However, first, I want to make sure you have the proper context.

Most of you should be familiar with the strategic planning tool that's referred to as SWOT analysis. SWOT stands for strengths, weaknesses, opportunities, and threats. It's used to synthesize data that relates to what a business is good at (strengths), where it's weak (or things that keep it from doing well), and the threats it faces from competitors and other influences. Opportunities (or options) represent the items that answer the question: "so what should we do about it?"

Most teams use a template to show items in each quadrant, but often, what's placed in a quadrant may not be backed-up by relevant, timely data. They use the tool to "fill-in-the-blanks" and brainstorm about opportunities. By the way, an opportunity really represents a possible future investment.

As an evolving strategic thinker, I encourage you to use SWOT as a true strategic synthesis tool to make sense of complex, interrelated data elements with the goal of asking "so what?" questions that require answers.

The Problem with Traditional SWOT

In my experience, traditional SWOT analysis produces lists that almost everyone already knows, obvious strengths, known weaknesses, visible opportunities, and clear threats. Teams are also heavily biased toward things they're predisposed to doing. The problem isn't that these observations are wrong; it's that they don't generate the strategic insights needed to make decisions that propel an organization forward.

Traditional SWOT treats each quadrant as isolated data collection rather than interconnected analysis. Teams fill in quadrants randomly, miss the connections between elements, and end up with static lists instead of dynamic insights.

THE ENHANCED SWOT TOOL

Strategic thinking transforms the SWOT model from a data and documents collection exercise to a strategic synthesis tool where opportunities or options represent areas in which an organization might invest to fulfill strategic goals. For each item in a quadrant of a SWOT model, you need to consider four issues: what it is, why it is important, what data or evidence supports it, and how each might connect to other quadrants.

There is one crucial nuance to applying strategic thinking to this model: address each quadrant in a specific sequence. Strengths first, then weaknesses followed by threats. This is important before you can start assessing opportunities. Notice the arrow in Figure 6.1 going counterclockwise from strengths through weaknesses and threats, ending at opportunities. Most people fill in quadrants randomly. This method requires addressing each quadrant in sequence and not identifying any opportunities until you've studied the interrelationships among strengths, weaknesses, and threats. This counterclockwise flow ensures your opportunities emerge from deep analysis rather than preconceived ideas.

If you don't have the needed data, you may need a plan to secure that data so your analysis is based on facts and evidence, not on guesses. I suggest viewing the model through the lenses of discovery (problem recognition), meaning (mindset), and solid strategic thinking.

For Strengths: Instead of asking "What are we good at?" ask:

- Which capabilities could help us create sustainable competitive advantage?
- What do we have that's valuable to customers and difficult for competitors to replicate?
- What strengths can we leverage to overcome a weakness or defend against a threat?

For Weaknesses: Focus on gaps that limit your ability to achieve desired outcomes or defend against threats. Use variance analysis to examine how these weaknesses are changing over time. Use competitive benchmarking and customer feedback to validate and quantify them. Refer to situations where a weakness caused an unintended outcome that harmed the business.

For Threats: Look beyond obvious competitive moves to systemic shifts that could undermine your business model. What market, technology, or regulatory changes could make your strengths irrelevant? The main idea is to surface vulnerabilities that emerge from inside the organization (e.g., technology or a staffing shortage) or from outside (e.g., economic indicators or new competitors).

For Opportunities: *Now, and only now,* identify opportunities that emerge from the intersection of your analysis. Where can your strengths be leveraged to address market needs? What capabilities could you build to defend against threats while creating new value?

Figure 6.1 Enhanced SWOT Model

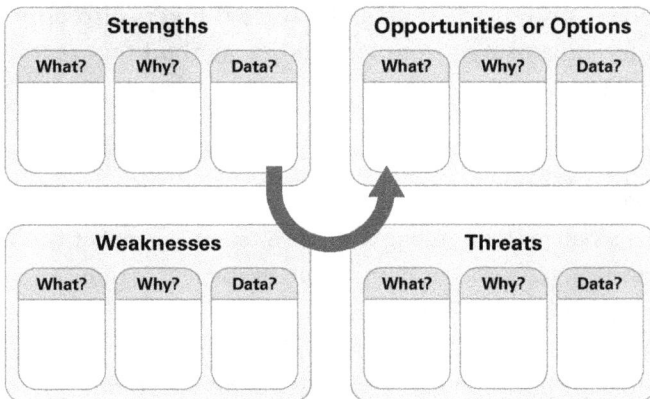

CROSS-SWOT ANALYSIS

The real strategic thinking power in SWOT comes from synthesis, or how the four quadrants connect and what those connections suggest about strategic direction. This is where strategic thinking truly differentiates itself.

After mapping connections and identifying themes using systems thinking, strategic thinkers ask the crucial question: "So what does this mean for our strategic direction?"
Look for patterns across quadrants:

- Which strengths could be leveraged to capture specific opportunities?
- Which weaknesses make you vulnerable to identified threats?
- How could addressing certain weaknesses unlock new opportunities?
- Which threats could be transformed into opportunities if you built the right capabilities?

This synthesis transforms SWOT from four separate lists into a coherent strategic narrative that guides option development and resource allocation decisions. However, understanding your current position and connections is only the first step. Strategic thinkers also approach goal setting differently, using a counterintuitive method that reveals insights traditional planning misses.

TRY THIS:
Bring a small team together with a whiteboard. Apply the enhanced SWOT process using your strategic thinking framework. Use pattern recognition to look for recurring themes across different data sources. Apply systems thinking to map relationships between quadrants. Ask better questions about what makes each item important and what evidence supports it. Consider time horizons for how each element might change over different periods. Leave opportunities blank until you've completed strengths, weaknesses, and threats, then draw connections between quadrants to reveal strategic insights.

START WITH DESIRED OUTCOMES AND WORK BACKWARD

Traditional goal setting starts with current reality and projects forward. Strategic thinkers examine data and information from multiple vantage points. That's why pattern mapping and zooming in and zooming out matter so much. When you use a linear approach, you may miss important signals. When you start with the goal and work backward, you get to see things differently. It's almost like checking your math in a linear equation. This approach applies the time horizon thinking from Chapter 2 systematically.

The most powerful strategic thinking question I use with clients is deceptively simple: What do you want to have happen? This isn't about vision statements or aspirational goals. It's about getting crystal clear on specific, measurable outcomes before getting lost in the complexity of how to achieve them.

The Three-Step Working Backward Process

Working backward transforms abstract strategic goals into concrete action pathways. Let me share how this three-step sequence makes working backward practical.

Step 1: Find Breakthrough Insights

This step transforms goal setting because it forces you to think about ends before means, outcomes before processes, and results before activities. Most strategic planning gets bogged down in analyzing current constraints and incrementally projecting forward. Working backward starts with clarity about desired destination and then figures out what path would get you there.

This demonstrates the mindset shift from Chapter 4, from "best practice" to "best for this situation." Instead of copying what others have done, you're designing a future that leverages your unique capabilities and market position.

Step 2: Work Across Time Horizons

Working backward through time horizons helps you understand what would need to be true at each stage. This applies the time horizon thinking in reverse chronological order.

You may start with goals to achieve two years into the future (market share, financial impact, market position). Then you work backward to understand what must be true about your business by the end of year 1 and by year 2. Consider which strategic initiatives (opportunities or options) can help you achieve those goals. Continue working backwards to identify quarterly actions to attain your goals.

Step 3: Find Capability Gaps

Working backward reveals capability gaps that must be closed to achieve desired outcomes. For each milestone, ask what capabilities, resources, or conditions don't exist today but would need to exist.

Each gap becomes a strategic problem worth solving. You can use the process from Chapter 5 to reframe, validate, generate options, and decide priorities. Identify which gaps are most critical and which could become bottlenecks that prevent progress toward other milestones. Apply systems thinking from Chapter 3 to understand how capabilities connect, which capabilities enable others, and which must be developed in sequence versus parallel.

Why Working Backward Works

This approach reveals strategic insights that forward-looking planning often misses. When you work backward, you discover capability gaps before they become crises. You identify resource requirements with enough lead time to secure them. You spot potential conflicts between different strategic initiatives early enough to resolve them.

Working backward also helps you test the feasibility of your strategic goals. If you can't map a credible pathway from current state to desired outcome, you either need to adjust your goals or invest in building the missing capabilities first.

Most importantly, working backward creates alignment around what success requires, not just what it would look like. Your team develops a shared understanding of the capabilities, resources, and sequence of moves needed to achieve strategic outcomes.

This working backward analysis becomes the foundation for developing strategic options that creates a solid connection to your desired outcomes rather than generic best practices that may or may not advance your strategic goals. Now you're ready to build your portfolio of strategic options.

Pick a specific strategic outcome you want to achieve in 18-24 months. Define the breakthrough insight that would need to be fundamentally different about your business? Work backward through 6-month milestones, asking what capabilities, resources, and conditions must exist. Apply systems thinking to identify which capability gaps are connected and must be developed in sequence. Use your learning mindset to examine what assumptions you're making about timeline and feasibility.

STRATEGIC OPTIONS: BUILDING YOUR PORTFOLIO

From my standpoint, it's easier to identify strategic options or opportunities when you've assessed your baseline (which always shifts), synthesized your data with the enhanced SWOT, and worked backward.

This is where I see the biggest difference between strategic thinkers and traditional planners. Most teams generate strategic options through brainstorming sessions that produce familiar ideas: "We should expand into new markets," "We need to improve our technology," and "Let's focus on customer experience." These aren't necessarily wrong, but they're generic responses that could apply to almost any business.

Strategic thinking transforms option development from brainstorming generic strategic goals to generating specific options grounded in deep understanding of your business reality and market possibilities. This applies the strategic mindset shift (Chapter 4) from "best practice" to "best for this situation," directly to strategic goal setting and strategy formulation.

Three Categories of Strategic Options

In my workshops I've found that strategic options typically fall into three categories, each serving different strategic purposes:

1. *Known necessities:* These are things you know will likely need to be done. You may have been talking about these with your team already. However, you've rationalized what these items might be, based on your baseline assessment and SWOT insights. They emerge from clear gaps in your current position or obvious requirements for your desired outcomes. While they may seem routine, strategic thinking helps you approach them in ways that build capabilities rather than just checking boxes.

2. *Position-fortifying options:* These may include product enhancements, service improvements, or capability investments that strengthen your competitive position. They build on strengths identified in your SWOT analysis and address weaknesses that make you vulnerable. These options don't necessarily transform your business, but they can fortify your position and help the company to become more competitive.

3. *Strategic investments:* These are longer-term capability building and transformational moves that create new competitive advantages or prepare you for market evolution. These can include new innovative offers, structural changes to the business, and new equipment or facilities. These often emerge from a working backward analysis when you realize your desired outcomes require capabilities you don't currently possess.

With your strategic options categorized, your ability to think strategically contributes to a more cohesive evaluation across multiple dimensions rather than a single criterion, such as return on investment (ROI), market size, or revenue potential. Now I'd like to take the multi-lens approach I mentioned in Chapter 3 and show you how to view strategic options. This multi-lens evaluation prevents you from choosing options that look good in one dimension but create problems in others. Figure 6.2 shows how the five lenses might work together.

Figure 6.2 Multi-lens Evaluation

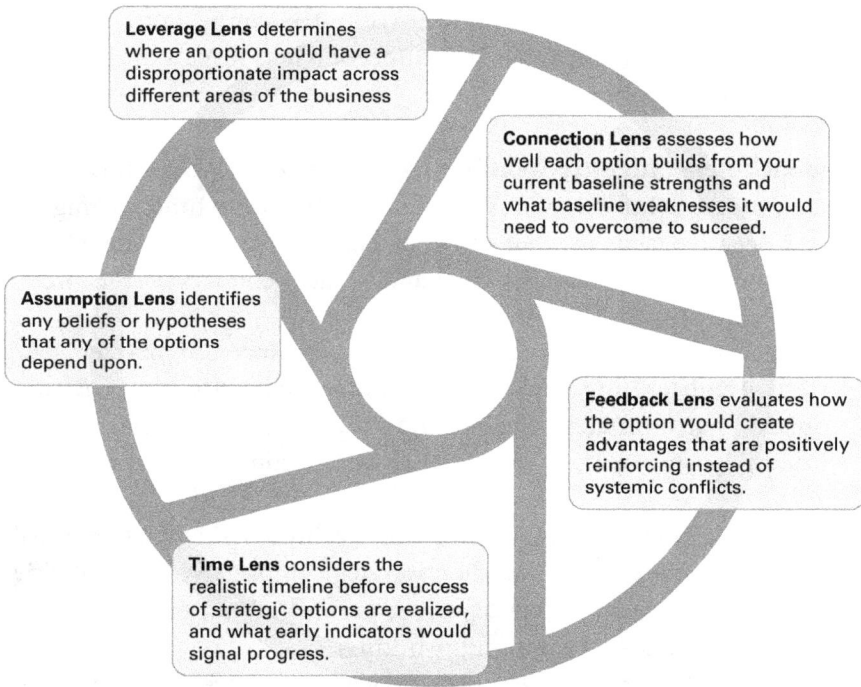

Leverage Lens determines where an option could have a disproportionate impact across different areas of the business

Connection Lens assesses how well each option builds from your current baseline strengths and what baseline weaknesses it would need to overcome to succeed.

Assumption Lens identifies any beliefs or hypotheses that any of the options depend upon.

Feedback Lens evaluates how the option would create advantages that are positively reinforcing instead of systemic conflicts.

Time Lens considers the realistic timeline before success of strategic options are realized, and what early indicators would signal progress.

MULTI-LENS STRATEGIC EVALUATION: SEEING OPTIONS THROUGH SYSTEMS THINKING

Most teams evaluate strategic options using single criteria such as ROI, market size, or competitive advantage. Strategic thinkers apply the five systems lenses from Chapter 3 to see how options perform across multiple dimensions simultaneously. This prevents choosing options that look brilliant in isolation but create unintended consequences when implemented.

Applying the Five Lenses Systematically

Connection Lens: Map Strategic Interdependencies. Ask: How does this option connect to our current baseline strengths and weaknesses? What other business functions would it affect? Would it create positive reinforcement with existing capabilities or reveal new dependencies?

For example, a customer service enhancement option might strengthen your competitive position (positive connection) while requiring significant technology upgrades (resource connection) and new hiring (capability connection). The connection lens reveals the full scope of what "simple" options require.

Feedback Lens: Identify Reinforcing Effects. Ask: Would this option create positive reinforcing loops that strengthen over time, or might it trigger negative feedback that works against your strategic intent? How would competitors likely respond, and what secondary effects would their responses create?

A pricing strategy option might initially increase market share (positive feedback) but could trigger competitive price wars (negative feedback) that ultimately damage industry profitability. The feedback lens helps you anticipate these dynamics before committing resources.

Time Lens: Understand Implementation Realities. Ask: What's the realistic timeline before this option shows strategic impact? What capabilities must be developed in sequence versus parallel? What early indicators would signal whether the option is working?

Many strategic options fail because leaders underestimate development timelines or expect results before the option has had time to work. The time lens prevents both premature abandonment and unrealistic expectations.

Assumption Lens: Surface Hidden Dependencies. Ask: What assumptions about markets, customers, competitors, or internal capabilities does this option depend on? What would happen if key assumptions proved incorrect? How could we test critical assumptions before full commitment?

Strategic options often fail because they're built on outdated assumptions about customer needs, competitive responses, or internal capabilities. The assumption lens makes these dependencies explicit so you can validate or adjust them.

Leverage Lens: Assess Strategic Impact. Ask: Where would this option create disproportionate impact across different areas of the business?

Could it address multiple strategic challenges simultaneously? Would it build capabilities that enable future strategic options?

The highest-value strategic options often have leverage effects because they solve immediate problems while building capabilities for future challenges. The leverage lens helps you identify options that create compound strategic benefits.

Integrating Multi-Lens Insights

After applying all five lenses, look for patterns across your analysis:

Strategic Coherence Check: Do the lenses reveal that this option reinforces your strategic direction, or do they expose conflicts and contradictions? Options that perform well across multiple lenses typically create stronger, more sustainable outcomes.

Risk-Benefit Integration: Use insights from the assumption and time lenses to understand what could go wrong and when you'd know. Combine this with connection and leverage lens insights to assess whether potential benefits justify identified risks.

Implementation Sequencing: Apply feedback and time lens insights to understand not just whether to pursue an option, but how and when to implement it for maximum strategic effect.

The Strategic Thinking Integration

This multi-lens approach directly applies your strategic thinking capabilities:

- Better questions (Chapter 2): Each lens generates specific questions that reveal insights you'd miss with single-criterion evaluation
- Pattern recognition (Chapter 2): Looking across lenses helps you spot patterns in how different options might perform
- Systems awareness (Chapter 3): The lenses ensure you're evaluating options as system interventions, not isolated changes
- Learning mindset (Chapter 4): Multi-lens evaluation builds in the assumption testing and reality-checking that prevents strategic surprise

Beyond Binary Decisions

The most powerful insight from multi-lens evaluation often isn't which option to choose, but how to modify options to perform better across multiple dimensions. You might discover that combining elements from different options creates a stronger strategic approach, or that sequencing options differently reduces risks while maintaining benefits. In decision-making, we refer to this as "combining options." However, when there are options that capitalize on all five lenses, there's a good chance you'll create something that can create real competitive advantage!

When viewing options through these lenses, you can evaluate how each option addresses insights from your SWOT assessment. Do the options leverage strengths and capture opportunities while addressing weaknesses or defending against threats? You determine how the option advances progress toward working backward milestones and addresses working backward capability requirements.

The systems impact analysis is crucial here. How would this or any option affect other parts of your business system? What positive or negative ripple effects could it create? This prevents you from optimizing one part of your business while accidentally breaking another part.

TRY THIS:

Categorize your current strategic options using the three-category framework. For each option, ask: Is this a known necessity, position-fortifying move, or strategic investment? Then apply the five lenses from Chapter 3 to assess systems impact. Look for options that score well across multiple dimensions rather than just excelling in one area.

With your strategic options categorized, strategic thinking enables more comprehensive evaluation across multiple dimensions rather than single criteria like ROI or market size. The multi-lens approach from Chapter 3 prevents you from choosing options that look good in one dimension but create problems in others.

PORTFOLIO THINKING FOR STRATEGIC OPTIONS

Strategic thinkers don't just evaluate options individually; they think about how options work together and sequence over time to create mutually reinforcing strategic direction. This portfolio approach is essential because no single strategic option is likely to achieve your desired outcomes by itself. This systematically applies the time horizon thinking discussed in Chapter 3 to option portfolio design.

Your three categories of options naturally align with different time horizons and risk profiles, as shown in Figure 6.4. Known necessities typically deliver results within 6–12 months with lower risk. Position-fortifying options usually require 12–18 months and carry moderate risk. Strategic investments often need 18+ months and involve higher risk but create transformational capability.

Figure 6.4 Portfolio Approach for Strategic Options

6–12 months
Known necessities
- Lower risk
- Address clear gaps
- Predictable returns
- Build capabilities

18–12 months
Fortifying position
- Moderate risk
- Strengthens competitive position
- Steady measurable improvement

18+ months
Strategic investments
- Higher risk
- Transformational potential
- Advance competitive advantage

The optimal portfolio balance depends on your competitive context and strategic urgency. Companies in stable markets with strong positions can emphasize known necessities and position-fortifying options while making selective strategic investments. Organizations facing rapid market change or competitive threats may need to weight their portfolio toward strategic investments despite higher risk.

The key is creating reinforcing effects across time horizons where near-term successes fund longer-term investments, and capability building enables both defensive and offensive strategic moves. Without this

portfolio perspective, you end up with competing initiatives that frag-
ment resources and slow strategic progress

Creating a balanced portfolio of strategic options is only half the
battle. The real test of strategic thinking comes in execution: transforming how you allocate resources to support these strategic priorities rather
than historical patterns.

SETTING UP PATHWAYS TO EXECUTION: FROM RESOURCE ALLOCATION TO STRATEGIC ENABLEMENT

In my corporate life, the strategic planning cycle seemed to resemble a
budgeting cycle; when the strategy was ready for execution, the budgets
were released, and the functional departments got to work. This may
work in complex companies, such as communications firms, banks, large
industrial firms, and others where market momentum and the sheer size
of the customer base can keep things moving with steady, predictable
cash inflows. These firms do strategize intensely, but this tends to happen
very high up in the organization or within the business units.

Recognize that strategic thinking provides the interventions needed
to transform resource allocation from budgeting to strategic enablement.
Instead of allocating resources based on historical patterns or departmental requests, resources are allocated based on strategic option priorities,
working backward requirements, and capability development needs.

Remember how variance analysis reveals hidden stories about what's
really happening in your business? The same principle applies to resource
allocation. Traditional budgeting looks at resource allocation variances.
You might ask, "Did we spend more or less than planned?" However, this
treats variances as isolated events. Strategic thinking asks a deeper question: What do our resource allocation patterns tell us about our actual
strategic priorities versus our stated ones?

I've seen this disconnect countless times. A company will declare that
innovation is their top strategic priority, then allocate 80 percent of their
resources to maintaining current operations and only 5 percent to developing new capabilities. The variance between the stated strategy and the
actual resource allocation reveals what the organization really prioritizes,

regardless of what the strategic plan says. Instead of defending current resource allocation patterns, strategic thinking treats resource allocation variances as information about strategic alignment and organizational behavior.

The way I see it, financial and human resources aren't just expenses to be managed; they're investments in capabilities that enable strategic options. When you understand this, resource allocation becomes strategic decision-making rather than budget administration. You ask questions such as What percentage of our current resources support our highest-priority strategic options? How much capacity is trapped in activities that don't advance working backward milestones? Where are resources allocated based on historical patterns rather than strategic priorities? You examine resource allocation effectiveness by looking at variance patterns using the detective work approach. Are we consistently underinvesting in capabilities that strategic options require? Are we overinvesting in areas that don't create strategic value? What do prior period budget variances tell us about the accuracy of our strategic assumptions?

One client company discovered through resource allocation variance analysis that they were systematically underfunding customer success initiatives while overfunding product development. The variance pattern revealed a strategic disconnect: they were prioritizing new feature development over ensuring existing customers achieved value. This insight led to a fundamental rebalancing that improved customer retention by 35 percent and reduced the need for new customer acquisition.

The main idea is to decide which resource investments build capabilities needed for multiple strategic options or examine how you can create resource flexibility that adapts as your baseline shifts and strategic options evolve. This requires thinking about resources as a portfolio, not just individual budget line items.

Consider partnerships, alliances, or outsourcing arrangements that could provide capabilities without permanent resource commitments. These resource options become especially valuable in uncertain environments where you need flexibility to adapt as conditions change. You're not just allocating resources; you're creating options for the allocation of resources that are aligned with the strategic intent of the firm.

TRY THIS:

- Step 1: Strategic Priority Audit If feasible, conduct a strategic priority audit of your company's current resource allocations. First, list your top 3 stated strategic priorities. Then, calculate what percentage of your total resources (budget and people) supports each priority. Look for the gaps between what you say matters and where your resources go. Use variance analysis from Chapter 5 to identify patterns. Probing questions: Are you underinvesting in certain capabilities? Do you have the right technology? How are other strategic projects performing (e.g., contributing to results)?

- Step 2: Resource Capability Mapping Create a resource capability map using the Connection Lens from Chapter 3. List your strategic options in one column and required capabilities in another. For each capability, identify: (1) current resource allocation, (2) strategic importance, (3) capability gaps. Look for capabilities that enable multiple strategic options. These deserve priority investment. Design resource options like partnerships or flexible arrangements that provide capabilities without permanent commitments.

But how do you know if your resource allocation is working? How do you track progress toward strategic outcomes while maintaining the flexibility to adapt as conditions change? This requires a fundamental shift in how you think about measurement.

KEY PERFORMANCE INDICATORS (KPIS): GUIDEPOSTS FOR STRATEGIC NAVIGATION

Strategic thinking transforms metrics from historical reporting to future-focused strategic navigation. Instead of simply measuring what happened, you measure progress toward strategic outcomes, early signals of baseline shifts, and leading indicators of strategic option success.

Leading vs. Lagging Indicators

Think of it this way: your car's speedometer tells you how fast you were going (lagging indicator), but looking ahead at the road conditions tells you whether you'll need to slow down or speed up (leading indicator).

Lagging indicators tell you what already happened. They can include revenue last quarter, customer satisfaction last month, market share last year, and customer churn rate last month. These indicators are like looking in your rearview mirror while driving, but they don't help you navigate what's coming.

Leading indicators predict what's likely to happen before it shows up in your traditional reports. They're like looking through your windshield to see what's ahead. Leading indicators may include declining engagement scores that predict future churn and sales activity metrics that predict pipeline quality changes. Another example: a drop in sales activity metrics serves as a leading indicator that predicts a near-term drop in revenue.

The goal is to spot patterns that give you three to six months' advance warning so you can adjust course before problems show up in your quarterly reports.

This detective thinking from Chapter 5 becomes essential for designing strategic key performance indicators (KPIs). You need to think like a detective, asking What early warning signals would indicate our strategic options are working or failing? What variance patterns historically preceded strategic successes or failures in our business? What combinations of small changes across different metrics suggest bigger strategic shifts developing?

BUILDING STRATEGIC METRICS THAT MATTER

Detective thinking helps you build metrics that matter for your strategic work. Instead of just tracking what happened, you can design metrics that reveal how your strategic foundation is changing and whether your strategic options will work as expected in the future.

Focus on What You Can Influence

You don't need enterprise-level authority to build strategic metrics. Start with metrics that matter for strategic options that relate to your role, projects, or products. The key is shifting from What did we accomplish? to What patterns suggest what's coming next? Here are three types of metrics you may wish to track:

1. *Strategic option progress indicators:* How well are your strategic initiatives performing against expectations? Instead of just measuring completion percentages, track early performance signals that predict success or failure. Are you building the capabilities you intended? Are you seeing the market response you expected?

2. *Baseline shift indicators:* How is your competitive position changing? Track variance patterns that historically predicted strategic challenges or opportunities in your business. Are your strengths getting stronger or weaker? Are your weaknesses becoming more problematic or less relevant?

3. *Working backward progress indicators:* Are you making progress toward your desired outcomes? Track milestone achievement rates, capability development progress, and market position improvements that support your working backward timeline.

As you progress, consider metrics that create insights when connected. For example, relate product quality indicators to complaints and returns, or examine rates of change in metrics that might signal problems across departments. You may see a drop in engineering headcount that leads to slower product introductions, which then negatively impacts revenue and profit.

To get ahead of the curve, build an early warning system that gives you three to six months' advance notice on strategic issues. Pick a few key metrics, track them over time, then use probing questions to understand what's happening and how to stay on track. These measurement capabilities complete your strategic thinking toolkit. But the real power emerges when all these elements work together, creating compound effects that transform both your strategic capabilities and your organization's intelligence.

INTEGRATION: HOW IT ALL WORKS TOGETHER

The real transformation happens through integration and compound effect. Each capability reinforces the others, creating strategic thinking power that exceeds the sum of individual tools. Better questions from Chapter 2 reveal system patterns across time horizons. Systems thinking from Chapter 3 improves problem recognition and solution design from Chapter 5. Strategic mindset from Chapter 4 enables effective strategy formulation under uncertainty.

This compound effect accelerates over time. The more you apply strategic thinking, the more natural it becomes. The more you see opportunities to apply it, the better your results become. The better your results, the more confident you become in strategic thinking approaches.

Organizations that master this integration create compound advantages that build over time. They spot opportunities and threats earlier through continuous baseline pattern recognition. They make better strategic choices because decisions are grounded in deeper business insights. They adapt faster because they've built learning into their strategic processes from the beginning.

Most importantly, they create organizational intelligence that gets smarter with every challenge, turning strategy formulation from an annual planning event into continuous capability development that builds sustainable competitive advantage.

YOUR STRATEGIC THINKING JOURNEY CONTINUES

You now have the foundation to create strategies that work in dynamic environments. The five habits, systems thinking, strategic mindset, and problem-solving capabilities all come together in the ongoing work of strategy formulation.

The transformation you've experienced through this book represents more than skill development. It's a fundamental shift in how you approach complex challenges, uncertain environments, and leadership opportunities. You started with five strategic thinking habits that seemed simple but proved powerful in application. You developed systems awareness that revealed hidden connections shaping business outcomes.

You cultivated a strategic mindset that embraces uncertainty as information rather than threat. You mastered problem-solving approaches that address root causes rather than symptoms.

Strategy formulation ends where it began: with thinking. But now it's not just individual thinking, it's organizational intelligence. When strategic thinking becomes embedded in how people see, decide, and act, strategy becomes not just a plan, but a way of life.

The leaders who will thrive in the coming decades won't just be those who know more; they'll be those who think better. They'll be strategic thinkers who can navigate complexity, adapt to uncertainty, and create value through intelligent decision-making.

You're ready to be one of those leaders. The question is how you'll use these capabilities to create strategies that thrive in complexity and uncertainty.

Start applying them now. Your next strategic decision is an opportunity to demonstrate the power of strategic thinking in action.

SUMMARY

This chapter shows how strategic thinking can transform how you approach the fundamental process of strategy formulation. Instead of mechanically following planning frameworks and linear models, you now approach each element with the analytical depth and systems awareness that helps you solidify goals and identify strategies that help you put your company ahead of the competition. To summarize, here are the six key transformational tools I recommended in this chapter.

1. Dynamic Baseline Thinking. Treat your business position in the market as a constantly moving line like a line of scrimmage on a football field, influenced by an ongoing inflow of data. Because of this, you can leverage AI to track patterns and connections at scale. These will influence your perspectives so you can figure out "what's next?"

2. Enhanced SWOT Analysis. Instead of haphazardly filling in the quadrants, focus on the synthesis of data quadrant-by-quadrant, sequentially (strengths first, weaknesses second, and threats

third). Make sure you focus on cross-quadrant connections to clarify strategic options.

3. **Working Backward Planning.** Start with desired outcomes and work backward through time horizons to identify capability gaps and breakthrough insights that forward planning misses. When you begin with the end in mind, it may be easier to filter options to ones that might have the greatest impact on results.

4. **Portfolio Strategic Options.** Your company's executives allocate funds to various investments to help fulfill its strategic intent. I suggest you become familiar with these. You can support their goals when you organize strategic investments associated with your work when you organize investment options into three categories: known necessities, position-fortifying, and strategic investments that produce optimal outcomes over the long run.

5. **Strategic Resource Enablement.** Allocate resources based on strategic priorities, not historical patterns. Use variance analysis to ensure alignment between stated and actual priorities.

6. **Future-Focused Metrics.** Design KPIs that predict what's coming (leading indicators) rather than just report what happened (lagging indicators).

The power lies not in any single tool, but in how these capabilities integrate to create organizational intelligence that adapts and strengthens with every strategic challenge.

A PERSONAL NOTE ON STRATEGIC THINKING

After two decades of working with leaders across industries, I've witnessed something troubling: brilliant, capable professionals trapped in tactical thinking patterns that limit their impact and stunt their growth. They work incredibly hard, solve problems efficiently, and deliver results. Yet, they remain frustrated by their inability to break through to the next level.

The breakthrough isn't about working harder or getting smarter. It's about fundamentally changing how you think about complex challenges.

I wrote this book because strategic thinking should not be reserved for the C-suite or remain a mysterious capability that some people "just have." It's a learnable discipline with specific habits, frameworks, and mindsets that anyone can develop. I've seen mid-level managers transform into strategic leaders, entrepreneurs build more resilient businesses, and entire teams elevate their problem-solving capability by applying these principles.

My motivation is deeply personal. I want to democratize strategic thinking. I'm offering you tools that traditionally only came through years of expensive trial and error. Every exercise, framework, and assessment in these pages emerged from real client work, tested with thousands of professionals facing genuine business challenges.

This represents a tremendous opportunity for your professional growth because strategic thinking capability compounds exponentially. Once you develop pattern recognition, systems awareness, and strategic problem-solving skills, they enhance everything else you do. You'll see opportunities others miss, solve problems others can't, and influence outcomes in ways that get you noticed for the right reasons.

Most importantly, this expands your field of view. Instead of being reactive to events, you'll anticipate them. Instead of optimizing within constraints, you'll question the constraints themselves. Instead of competing in existing spaces, you'll recognize where new spaces are emerging.

The gap between tactical and strategic thinking is the difference between managing your current role well and creating your next opportunity. My hope is that these capabilities become so natural for you that strategic thinking stops being something you do occasionally and becomes how you approach every meaningful challenge.

The leaders who shape the future don't just think differently, they think strategically. Now you can too.

GLOSSARY OF KEY TERMS

Assumption Lens - One of the five systems thinking lenses that reveals the often-invisible beliefs and mental models that drive system behavior. Used to surface hidden dependencies that could undermine strategic initiatives.

AI Strategic Partnership - The collaborative approach where artificial intelligence handles pattern recognition and data processing at scale while humans provide strategic context, interpretation, and judgment.

Balancing Loops - Feedback loops that resist change and try to maintain system equilibrium. When a system pushes back against interventions to restore balance.

Best for the Situation - A strategic mindset shift away from applying generic "best practices" toward developing contextual solutions that fit specific organizational circumstances, culture, and constraints.

Better Questions - One of the five strategic thinking habits focused on asking probing, open-ended questions that reveal insights rather than confirming existing assumptions.

Business Acumen - The portfolio of skills, behaviors, and capabilities needed to support an organization in achieving its financial and strategic goals.

Business Ecosystem - The network of interconnected entities including customers, suppliers, distributors, regulators, and other stakeholders that influence organizational performance and strategic options.

C

Change Multiplier - A systems thinking tool that identifies intervention points where small changes create significant positive ripple effects across multiple organizational levels.

Connection Lens - One of the five systems thinking lenses that maps how different people, teams, processes, and outcomes interconnect, revealing hidden bottlenecks and leverage points.

Connection Web - A visual mapping tool that shows how different elements of a business challenge connect to each other, helping identify system patterns and intervention opportunities.

Connected Variance Analysis - The practice of examining multiple business metrics simultaneously to understand the stories that patterns across different data streams tell about underlying system dynamics.

Contextual Thinking - The ability to adapt solutions based on specific situational factors rather than applying universal approaches regardless of circumstances.

Creative Thinking - The cognitive ability to envision what could be and generate novel ideas for solving problems, complementing critical thinking in strategic analysis.

Critical Thinking - The process of thinking about your thinking while you're thinking to make your thinking better. Examining information objectively to uncover deeper meaning and contradictions.

Cross-Domain Pattern Discovery - The ability to spot connections and solutions by recognizing patterns across different industries, functions, or business domains.

D

Decision Matrix - A systematic tool for evaluating multiple options against weighted criteria that matter most for a specific situation, making trade-offs visible and decisions more objective.

Dynamic Baseline - The recognition that an organization's strategic position (like a line of scrimmage in football) constantly shifts based on market conditions, competitive actions, and internal changes.

E

Enhanced SWOT Analysis - A strategic synthesis tool that examines strengths, weaknesses, threats, and opportunities in sequence, focusing on cross-quadrant connections to reveal strategic insights rather than creating isolated lists.

Emergence - A systems principle where unexpected events or behaviors arise from complex interactions between system elements, often creating outcomes that couldn't be predicted from examining individual parts.

Environmental Cues - External signals or reminders that prompt strategic thinking behaviors and help maintain focus on long-term priorities despite immediate pressures.

F

Feedback Lens - One of the five systems thinking lenses that identifies reinforcing and balancing loops that drive system behavior over time.

Feedback Loop - A cycle where information or results from a system's output circle back to influence future inputs and actions, creating either reinforcing or balancing effects.

Feedforward Mechanisms - Systems that help organizations anticipate and prepare for future challenges rather than just learning from past results.

I

Important But Not Urgent - The strategic sweet spot where most significant problems begin. Issues that deserve attention before they become crises but don't yet demand immediate action.

Influence Conditions - A strategic mindset shift from trying to control specific outcomes toward shaping the conditions, culture, and processes that make desired outcomes more likely.

Innovation Connection - The relationship between strategic thinking capabilities (pattern recognition, systems awareness, creative problem-solving) and the ability to envision breakthrough possibilities.

K

Known Necessities - Strategic options that emerge from clear gaps in current position or obvious requirements for desired outcomes. Things you know likely need to be done based on baseline assessment.

L

Lagging Indicators - Metrics that report what already happened, like revenue last quarter or customer satisfaction last month. Historical data that confirms results but doesn't predict future performance.

Latticework of Mental Models - Charlie Munger's concept of building interconnected mental frameworks from multiple disciplines to better understand and navigate complex situations.

Leading Indicators - Metrics that predict what's likely to happen before it shows up in traditional reports, providing three to six months advance warning of strategic issues.

Learning Debrief - A systematic reflection process after significant meetings, decisions, or project milestones that captures insights while they're fresh and builds organizational intelligence.

Learning Mindset - One of the five strategic thinking habits. The approach of treating every situation as a learning opportunity rather than rushing to provide solutions.

Leverage Lens - One of the five systems thinking lenses that finds intervention points where small changes create disproportionate impact across different areas of the business.

Linear Thinking - The assumption that cause and effect follow simple, predictable chains (A causes B causes C) rather than recognizing complex, interconnected system dynamics.

Local Optimization Trap - The systems thinking error of improving one part of an organization while accidentally creating bigger problems elsewhere in the system.

Loop Tracker - A systems thinking tool that identifies feedback loops to determine whether they're making problems worse (reinforcing) or trying to maintain balance (balancing).

M

Mental Models - Internal frameworks or lenses through which you view and make sense of situations. The cognitive structures that shape how you interpret information and make decisions.

Mindset Lock - When previously successful mental models become rigid assumptions that prevent adaptation to changing circumstances.

Multi-Lens Evaluation - The practice of assessing strategic options through all five systems thinking lenses simultaneously to understand performance across multiple dimensions.

N

Nonlinearity - A systems principle where small changes can create disproportionate effects throughout an organization, and where cause-and-effect relationships are complex rather than direct.

O

Organizational Intelligence - The collective capability of an organization to sense, interpret, and respond to complex challenges in ways that build competitive advantage over time.

Organizational Sensing - The ability to detect signals that reveal how an organization really works versus how it's supposed to work according to formal structures.

P

Paleontologist's Approach - The strategic method of making good decisions with incomplete information by clearly distinguishing between verified facts ("fossils") and reasonable assumptions ("clay models").

Pattern Patience - The discipline of sitting with observations without rushing to conclusions, forcing consideration of multiple explanations before settling on the most obvious one.

Pattern Recognition - One of the five strategic thinking habits. The ability to connect seemingly unrelated observations across different business domains to gain new insights.

Position-Fortifying Options - Strategic investments that strengthen competitive position by building on identified strengths and addressing vulnerabilities that make you susceptible to threats.

Pre-Mortem Technique - A strategic thinking exercise where teams imagine a strategy has failed spectacularly and work backward to identify what assumptions were incorrect and what was missed.

Problem Iceberg - A four-level analysis tool that reveals what's really driving problems by examining events, patterns, structures, and mental models beneath surface symptoms.

Process Centricity - The organizational tendency to focus on rigid adherence to steps and procedures rather than adaptive thinking and creative problem-solving.

Q

Quick Fix Trap - The tendency to focus on symptoms rather than root causes, leading to recurring problems that consume resources without creating lasting solutions.

R

Reinforcing Loops - Feedback loops that amplify change, where an action creates conditions that intensify the original action. Can create virtuous cycles of success or vicious cycles of failure.

Ripple Effects - The secondary and tertiary consequences that spread throughout an organizational system when changes are made in one area.

S

Signal-to-Noise Ratio - The ability to distinguish between strategically significant patterns ("signals") and routine operational variations ("noise") that don't require strategic attention.

Situational Intelligence - The ability to read the unique characteristics of your environment and adapt solutions accordingly based on context, culture, and competitive position.

Step Zero - The problem recognition phase that occurs before traditional problem-solving begins. The capability to spot strategic issues while they're still manageable.

Strategic Investment Options - Longer-term capability building and transformational moves that create new competitive advantages or prepare for market evolution.

Strategic Intuition - The accumulated wisdom from years of experience that processes patterns faster than conscious analysis, serving as a quality check after rigorous analytical work.

Strategic Mindset - The mental foundation that embraces uncertainty as information, focuses on learning rather than knowing, and seeks contextual solutions rather than best practices.

Strategic Problem-Solving - The systematic approach to recognizing problems worth solving, reframing them through systems thinking, and designing solutions that address root causes while strengthening organizational capabilities.

Strategic Thinking - A mental process used to make sense of dynamic data and disparate observations to expose patterns, establish meaningful goals, determine appropriate courses of action, and continually refine implementation through feedback across different time horizons.

Strategic Thinking Mental Architecture - The five interconnected capabilities that separate strategic thinkers: five habits of strategic thinkers, systems thinking lenses, mindset shifts, problem-solving processes, and strategy formulation approaches.

Systems Thinking - The ability to see beyond isolated events and symptoms to understand the web of relationships and interdependencies that shape outcomes in complex organizations.

T

Temporal Discipline - The strategic practice of considering implications of choices over various time horizons rather than focusing solely on immediate results.

Temporal Leverage - Understanding how timing multiplies impact. Recognizing that some decisions have disproportionate effects over time.

Temporal Traps - Decisions that solve immediate problems but create more profound long-term issues by failing to consider consequences across time horizons.

Thinking Partner - A trusted colleague or advisor who helps you test strategic thinking, challenge assumptions, and explore different perspectives on complex challenges.

Time Lens - One of the five systems thinking lenses that reveals how delays distort cause and effect, helping leaders understand realistic timelines for strategic initiatives.

Time Horizon Thinking - One of the five strategic thinking habits. The discipline of considering immediate, medium-term, and long-term implications of decisions to prevent short-term optimization at long-term expense.

Timeline Reality Check - A systems thinking tool that prevents abandoning good strategies too early or sticking with bad ones too long by mapping realistic timelines between actions and results.

U

Urgency Addiction - The organizational conditioning to respond with equal intensity to all situations, leading to reactive decision-making and inability to distinguish between urgent and important issues.

Urgency Importance Matrix - A prioritization framework that helps distinguish between urgent problems (requiring immediate attention) and important problems (requiring strategic attention) to focus energy on the highest-impact activities.

V

Variance Analysis - The detective work of examining gaps between planned and actual performance to understand what stories these patterns tell about underlying business system dynamics.

Variance Patterns - The recurring themes and connections that emerge when examining multiple business metrics over time, revealing systemic issues that require strategic attention.

W

Working Backward - A strategic planning approach that starts with desired outcomes and works backward through time horizons to identify capability gaps and breakthrough insights that forward planning often misses.

Z

Zoom In and Zoom Out - One of the five strategic thinking habits. The ability to fluidly shift between detailed analysis and big-picture context to make balanced decisions informed by specifics and aligned with broader objectives.